Lecture Notes
in Economics and
Mathematical Systems

Managing Editors: M. Beckmann and W. Krelle

333

Nico Dellaert

AF866755

Production to Order

Models and Rules for Production Planning

Springer-Verlag
Berlin Heidelberg New York London Paris Tokyo Hong Kong

Editorial Board

H. Albach M. Beckmann (Managing Editor) P. Dhrymes
G. Fandel G. Feichtinger J. Green W. Hildenbrand W. Krelle (Managing Editor)
H.P. Künzi K. Ritter R. Sato U. Schittko P. Schönfeld R. Selten

Managing Editors

Prof. Dr. M. Beckmann
Brown University
Providence, RI 02912, USA

Prof. Dr. W. Krelle
Institut für Gesellschafts- und Wirtschaftswissenschaften
der Universität Bonn
Adenauerallee 24–42, D-5300 Bonn, FRG

Author

Dr. ir. Nico Dellaert
Erasmus University, Econometric Institute
P.O. Box 1738, 3000 DR Rotterdam, The Netherlands

ISBN-13:978-3-540-51309-4 e-ISBN-13:978-3-642-46672-4
DOI: 10.1007/978-3-642-46672-4

This work is subject to copyright. All rights are reserved, whether the whole or part of the material is concerned, specifically the rights of translation, reprinting, re-use of illustrations, recitation, broadcasting, reproduction on microfilms or in other ways, and storage in data banks. Duplication of this publication or parts thereof is only permitted under the provisions of the German Copyright Law of September 9, 1965, in its version of June 24, 1985, and a copyright fee must always be paid. Violations fall under the prosecution act of the German Copyright Law.

© Springer-Verlag Berlin Heidelberg 1989

Printing and binding: Druckhaus Beltz, Hemsbach/Bergstr.
2847/3140-543210

To Marie-José and Babbis

Table of Contents

Chapter 1

INTRODUCTION

1.1. Production to order

A type of situation that occurs frequently in process industry is the following one. A firm manufacturing a wide variety of products has a production process with one capacity bottle-neck with large set-up times. Demand for the products is highly uncertain and there is no possibility for having substantial stocks on hand. At the same time it is requested to deliver at short notice. Since no orders can be delivered from stock a due date has to be set for every order. This due date results from an agreement between the firm and the customer. By a proper scheduling of the production, these due dates should be met as close as possible, avoiding too many set-ups by clustering orders for the same product type. Clearly, in this kind of situation the control of the production and the way in which lead times can be determined is very important. Since these problems are not well covered by common planning methods, it would be interesting to investigate what kind of control frameworks can be set-up in situations like this.

One example of such a situation was found during a study concerning a company producing welded steel pipes (Dellaert and Wessels (1986)). These steel pipes are too voluminous to be stored in large amounts for a longer period. In fact no products are stocked at all, because the average amount that is ordered is quite substantial and the assortment of pipes that has to be manufactured is large and changing regularly. This company has clients with different priorities. The bottle-neck of the production process in this company appears to be the welding process. There are several parallel welding machines which are partially different. They all have to be reset every time another type of steel pipe is produced. To avoid too much loss of time for set-ups, one produces steel pipes belonging to the same type consecutively on the same machine. This clustering of course influences the production planning and the control of the lead times and the order acceptance considerably. Other examples of situations in which this combination of production to order, parallel machines with set-ups and a wide variety of products can be observed are generally found in process industry. Chemical industries, for instance, may use the same machines to produce different products. Every time another product is made on a machine, this machine has to be cleaned. Especially if the products are perishable and the demand is strongly varying and voluminous, it will make sense to produce to order. Similar problems can also be found in industries, like the truck industry and industries in telecommunication, where the products are manufactured in a way that is usually called assembly-to-order.

1.2. Description of the problem

According to a recent Swedish study (Mattsson et al. (1988)), about 80 percent of Swedish companies manufacture mainly on a make-to-order basis. Most of these companies have increased the degree of make-to-order production in the last seven years. Of course, production to order is only one element of a production situation: large vessels and aircraft will usually be manufactured on a make-to-order basis, but this equally holds for printed matter and for birthday cakes. Out of the many problem areas that manufacturing firms are faced with, we have chosen to study the problems concerning the production planning and the control of the lead times.

The form of the control rules with respect to the lead times depends on the situation. We may have a situation in which the firm uses rules to propose a lead time for every order, but we can also have a situation in which there are rules to accept or to refuse an order for which a certain lead time is asked by the client. According to the control rules for the production planning, the accepted orders are scheduled on the bottle-neck machine(s). In this monograph we want to find good control rules for situations in which the characteristics on market demand and on the production facilities are given. In particular, we are interested in the performance of these control rules in rather complex situations, for instance with different sorts of clients and a complex stochastic demand pattern and also different parallel machines, possibly with a partly flexible capacity. The set of control rules that can be used for the control in complex situations will be built up from the control rules for the typical simple situations. By considering the performance of the rules in these typical simple situations, we hope to get a better insight into the relations between the market demand and the production capabilities on the one hand and the control rules for planning production and lead times on the other. This knowledge can help us find good control rules for complex situations.

The quality of the control rules depends on the degree of satisfaction for both the firm and the clients. Of course, we will consider the same performance indices in the simple situations. This satisfaction, or performance, can be measured according to the wishes of the firm with regard to the demand, the smoothness of production, the amount of idle time and the number of set-ups. The performance can also be measured according to the wishes of the clients with regard to the length and accuracy of the lead times as well as the quality of the products. The wishes of the company and the wishes of the clients largely influence the rules that should be used for production planning or for controlling lead times. If, for instance, the firm wishes to produce the different types of products once every three weeks and the clients require a maximum delivery time of three weeks, a cyclic production rule seems obvious. However, if the firm considers a minimum level of demand necessary to justify a set-up and if the wishes of the clients concerning delivery times and lead times vary, some non-cyclic production rule has to be chosen.

1.3. The control rules

1.3.1. The control rules for the lead times

In order to avoid ambiguities, we first give some definitions. In Figure 1.1. these definitions will be illustrated. The arrival date of an order is the moment at which the order gets available for the production on the bottle-neck machine. The delivery date of an order is the moment at which the order is finished on the bottle-neck machine. The due date is the promised delivery date. The delivery time is the amount of time between the arrival date and the delivery date of an order. The lead time is the amount of time between the arrival date and the due date of an order. The residual lead time of an order at a certain moment is defined as the difference between this moment and the due date of the order. The residual lead time can have a negative value if the order has not been produced at the due date.

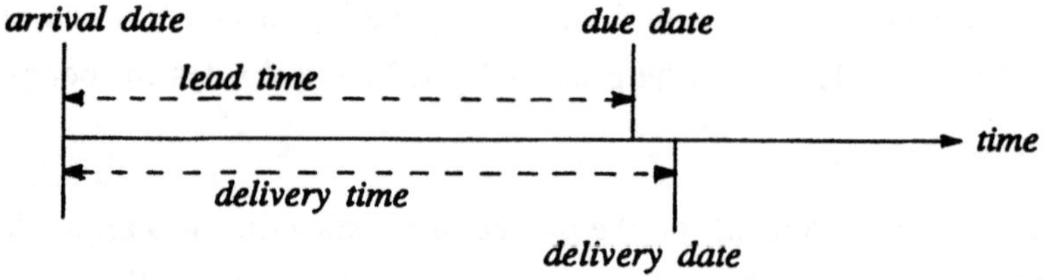

Figure 1.1 The process for an arbitrary order.

We can achieve a lead time for an order in three different ways. In the first place, the firm and the client may have a long-term agreement according to which the lead time of the order is fixed, the so-called fixed lead times. This implies that every time a client orders a certain item, the lead time for the order will be the same. In the second place, the firm can offer the client a lead time for the order, to which the client can agree or disagree, the so-called firm-initiated lead time. And in the third place, clients can ask the firm for a certain lead time for an order, the client-initiated lead time. The firm can refuse the order with this lead time or accept it.

In this monograph we will not consider any rules for making agreements about fixed lead times. In Dellaert (1987) this has been done for several examples. The lead time that the firm offers a client can depend on several factors:

1) the expected production date of the order based on the schedule at the arrival date;

2) the moment that gives the highest expected profit, based on the wishes of the client and consequences of production at that moment, both on the costs and on future orders.

If a client is not satisfied with the lead time that is offered, the order can be withdrawn or a new due date can be given, reconsidering the wishes of the client.

In the situation with client-initiated lead times the acceptance of an order with a certain required due date can also depend on the expected production date or on the expected profit of the order. In the expected profit not only the direct costs for holding, set-ups etc. should be considered, but also the effect of the order on future orders. In this study we will not consider situations with client-initiated lead times, because the modelling of these situations can be done in the same way as the modelling of the firm-initiated lead times.

Once the order and its due date are accepted, either by the firm or by the client, the due date has to be met as accurately as possible in order to avoid large costs and dissatisfied clients. It is the task of the production planning department to realise a delivery date that is close to the due date. In Chapter 2 we will consider the aspect of the difference between the due date and the delivery date, the so-called due date deviation, more closely.

1.3.2. The control rules for the production planning

The primary purpose of a production planning rule is to make sure that all orders are scheduled for production in such a way that the due dates are met as well as possible, while the set-up costs and overtime costs are kept low. Also possible negative effects on the delivery times of future orders should be avoided as much as possible. In the production planning the structure of the types, based on the set-up times, play an important role. Producing the same type too long leads to long delivery times for the other types, but on the other hand if the series is too small, too much capacity may be lost due to set-ups.

Regularly, decisions about the production have to be taken. These decisions can be taken at different levels in a hierarchical way. First we can do the capacity load planning and decide whether we will produce in a certain period or not. For this decision the demand forecasts at an aggregated level will be the most important element. For the periods in which we have decided to produce, we have to decide which type(s) will be produced. This decision will be called the type planning. The type that will be chosen may be the type that we have been working on most recently, especially if no set-up is needed in that case. It can also be the type for which the orders are the most urgent, possibly measured according to the demand forecasts. If we have chosen the type that will be produced, we have to select the orders to be produced and their sequence. This decision will be called order planning and it is based on the same elements as the choice of the type: the importance of the orders and the expected future demand.

The production planning can be done in the hierarchical way described above, but we can also combine two planning levels or even all three levels. If we do the scheduling in a hierarchical way, decisions about the work force level can be taken some weeks in advance. At a later moment, the sequence in which the types will be produced can be determined, for instance as soon as the number of orders is sufficient to justify a set-up for such a type. Again later, we must decide upon the orders within the type to produce. By taking the decisions about the production at different moments, it will be easier to organise the work, because a lot of uncertainty in the work can be removed. By combining two or three planning levels there is an increasing uncertainty about the products that will be produced at short notice, but we can gain flexibility by this combination. We can schedule the orders at the latest possible moment: if the previous order is finished we decide which order should be the following one to be produced. This way of scheduling is more flexible than the hierarchical way and therefore it offers more possibilities for short delivery times for urgent orders. Due to this flexibility the due dates can not be very accurate. The choice of the orders is based on the same aspects as the choice of the types: time lost due to a set-up, the urgency and importance of the order and the effect on future orders. Usually, making decisions about the orders is less complicated than making decisions about the types.

In this study we will concentrate us on the type planning. In a lot of practical situations the available capacity will be (almost) fixed. The capacity planning in such a situation can be considered as being a consequence of the type planning. The decision to produce a type will imply that we produce all orders for this type or a certain subset of the orders depending on the elements described above. Since the rules for the decisions about the orders will usually be quite simple, the type planning is the most interesting part of the planning.

1.4. The performance of the control rules

If we want to study the performance of a control rule in a certain situation, there are several elements of interest by which the performance can be measured. The demand process may depend on the acceptance of the proposed lead times. The length, the stability and the accuracy of the lead times can be important as well as the number of set-ups and the amount of extra capacity that is used. In the next chapter these elements are modelled as costs, enabling us to measure the performance by one single element: the total costs.

The first element of interest in the performance evaluation of a situation is the demand process. The demand can be realised in different ways. In the situation with client-initiated lead times the demand depends on the number of accepted orders by the firm. In the situation with firm-initiated lead times, the demand depends on the number of orders for which the proposed lead times are accepted by the clients. The firm and the clients can also make long-term agreements about deliveries over a longer period. The degree of uncertainty about the demand process depends of course on the fraction of orders for which long-term agreements have been made and on the choices for the delivery times. Since the resulting stochastic demand process determines the utility rate of the machines as well as the income for the company, the features of this process can be a very interesting part of the performance of a control rule.

The second interesting element is given by the lead times that are offered by the firm or that are accepted by the firm. The length and the stability of the firm-initiated lead times determine the demand process over a longer period, in combination of course with the realised delivery times. Most of the clients of a firm will stop ordering if they are not content with the long-term average of the length and the stability of the lead times. The same arguments will also be valid for the client-initiated lead times. A lot of orders with short lead times accepted by the firm may have an advertising effect, but if it is only orders with long lead times that are accepted, the clients may become discontent.

The third interesting performance measure is given by the accuracy of the lead times. The lead time that results from an agreement between the firm and the client should be met quite accurately. However, this will not always be possible, due to other, more urgent orders or due to reasons that have nothing to do with the production planning, such as machine breakdowns. Clients can react to the accuracy of the lead times in the same way as they did with the firm-initiated lead times: they can stop ordering if they are discontent or there can be an advertising effect if they are satisfied.

The fourth element in the evaluation of a control rule is the use of capacity. In a lot of situations we will be very interested in the percentage of time that is lost due to set-ups and the costs due to set-ups. The amount of required capacity may also be important, but more often the amount of extra capacity that is used will be an interesting result of a rule. The amount of extra capacity is especially interesting if the overtime costs are high and if the normal available capacity is tight.

1.5. Influences on the control rules

1.5.1. Influence of the market

At the level of strategic planning (see Anthony (1965)), one of the decisions a firm has to make is the decision about the group of clients it wants to serve. By focusing on the right groups of clients, the firm can offer these clients some long-term certainty. One of the most important decisions with a long-term effect is the decision whether to have client-initiated lead times, firm-initiated lead times or fixed lead times. These fixed lead times can be achieved sometimes with regular clients; i.e. clients that are always or very often supplied by the firm, who may have a good insight into their future need for products of the firm.

The choice of the clients and especially the choice of the way in which the lead times are determined is of course very important for the rule with respect to the lead times. A lot of elements of this control rule will be determined by these choices. Therefore the type of rule will be fixed for a long time.

In the situation of fixed lead times we have to make agreements, but also in other situations in which the clients have some insight into their future demand it might be recommendable to make some agreements with regular clients about the future demand and thus decrease the uncertainty for the firm and also for these clients. These agreements may contain different elements:

1) agreements in which the due dates, prices and the amounts to be delivered are fixed for a long term, in combination with fixed lead times;

2) agreements in which the client is obliged to order a fixed amount for a fixed price a number of times and in which the firm guarantees a maximum lead time for these orders (firm-initiated or client-initiated lead times);

3) agreements in which the client is obliged to order a certain amount during a period of 12 months, possibly in different quantities, and in which the firm gives some guarantees about the lead times (firm-initiated or client-initiated lead times).

As a result of these agreements there is less demand uncertainty, which makes the planning of the production easier. Because of the knowledge about the future production the firm will be able to calculate the effect of a certain price and a certain lead time for an order from an irregular client more precisely. Apart from the regularity of clients and the wishes about the length and stability of their lead times, another important aspect of the market, especially for the production planning, is given by the accuracy of the lead times clients are content with. For some clients it is important that the delivery date that is realised should not deviate more than for instance one day from the due date, while for other clients the critical margin may be one week, maybe because of a different pricing.

1.5.2. Influence of the production facilities

In this study we will consider production processes which have exactly one bottle-neck process. The bottle-neck part, in which this bottle-neck process takes place, may consist of one machine, several identical parallel machines or several parallel machines with slightly different features. Having a bottle-neck usually implies that this part of the production process is the most important part concerning the machine-availability, the variability of the delivery times and probably also concerning the variable part of the production costs. In our models we will consider no other production processes outside the bottle-neck process. Of course this largely simplifies the production planning problem.

Manufacturing a large variety of products makes a frequent change of production from one product to another inevitable. A change of production implies that a part of the bottle-neck machine has to be changed or has to be cleaned. Sometimes this may not take up much time, but large set-up or change-over times, with additional costs, are unavoidable. The products that are manufactured on the bottle-neck machine can be divided into groups, in such a way that between the products belonging to the same group only minor set-up times are necessary. We will refer to such a group as a type of product (see Bitran et al.(1981)). These groups will play an important role in the rules for production planning and therefore they will also be important in the rules for the lead times, because the due dates usually depend on the expected production moment. In many situations the influence of the minor set-ups is restricted to the sequence of the families of items. Therefore they will have little influence upon the delivery times. For this reason and because of the reduction of the complexity we will neglect the minor set-ups in our models.

1.6. The methods

Depending on the situation, we will model the demand process and the rules for planning production in such a way that the important features can be analysed. Some of the situations can be analysed in an exact way, but for others we will use an approximative analysis and in some situations the only analysing instrument will be a simulation study. It will often happen that an exact analysis is possible if the number of order states can be limited. The analysis of these small problems can help us develop intuition for the rules that should be used for larger problems, but the larger problems themselves have to be solved by approximative methods. The approximative analysis will of course be compared with the exact analysis, if possible. Otherwise, we can compare the approximative analysis with the simulation results. However, simulation studies of complex situations will be time consuming, especially if we want reliable results and if we consider many possible options for the decision variables. Therefore, we hope to determine the values of some of the decision variables by means of analysis.

Methods that will be used in this study are partially derived from well-known methods in the fields of production/inventory models, the queuing theory and Markov Decision Processes. The other methods that will be used, apart from simulation, are all based on the use of Markov chains. In a continuous review situation queuing models using Markov processes can be of much help. Queuing models assume that only the jobs or clients present in the system can be served, the main principle of production to order. Furthermore, all kinds of priority rules and distributions for demand and service times have been considered in literature. Therefore we will use a queuing model in a continuous review situation.

1.7. Overview of the text

In Chapter 2 we will model the situation in general and discuss some instruments for making decisions. Then we give types of control rules for the production planning and for the generation of the lead times and order acceptance. In every situation we can use a cyclic production rule for the production planning, although the performance of such a rule may be very bad. By using a production cycle in which the sequence of the types and the available capacity is fixed for a long period, proposing lead times for orders becomes very simple. Since this way of production planning is used a lot in practical situations, it will be considered in Chapter 3.

In Chapter 4 we study some of the interesting aspects in a rather isolated situation. Throughout this chapter we will assume that there are no capacity restrictions. This allows us to consider single type problems, because we can consider each type of product separately. This assumption can help us find good rules for more complex situations with capacity constraints. First we will study a situation with fixed lead times for different groups of clients. In this situation some analysis is possible. We will also develop a number of production rules that can be used in different situations. One of these rules is rather promising. This (x,T)–rule offers low average costs, it is easy to use and to analyse and it can be adapted to all kind of difficulties. We continue with a situation with fixed lead times, in which the level of demand depends on the lateness of previous orders, that is the demand level drops if many orders have been delivered too late. Afterwards, a situation is described in which the firm proposes lead times for the orders. The acceptance of the lead times happens according to a stochastic process.

A more complex situation will be considered in Chapter 5. We will consider a situation in which several types of products are produced on one machine, with a limited capacity. The capacity is fixed in one situation and can be extended by making overtime in another situation. In both situations the (x,T)–rule will be extended and compared with a more complex production rule, based on well-known methods. There will be different groups of clients with different fixed lead times. In some examples we will compare the two production rules with the cyclic production rules that have been described in Chapter 3.

In Chapter 6 we will again consider a situation with different types of products on one machine, with a limited capacity. By assuming some simple rules for the production planning, based on the (x,T)–rule, we can perform an approximative analysis. This analysis yields a distribution of the delivery times for orders of different types and priorities in the periodic review model. We will also consider a queuing model with exponentially distributed service and interarrival times. In this model, the approximative analysis yields average delivery times for orders of different types.

The insight into the different problem aspects will be combined in Chapter 7. In this chapter we shall consider the most complex situations, with orders with different priorities and for different types. Firm-initiated lead times are proposed for the orders, based on a preliminary production plan. There is a given probability that the customer withdraws an order if the proposed lead time is too long. The orders will be produced on several machines. Some types of products can be produced on one machine, other types on two machines. This complicates the production planning, but even in this situation we can use a simple production rule, based on the (x,T)–rule, in combination with some special extensions. Finally, we will give the conclusions of our study in Chapter 8.

Chapter 2

PRELIMINARIES

2.1. The situation and the model

The situation we will consider is that of a firm, possibly in the process industry, manufacturing a wide variety of products on a make-to-order basis. We are particularly interested in those production processes which have exactly one bottleneck, not only because this situation is quite common, but also because it is the situation that can be analysed best. In our models we will only consider the bottleneck process and exclude the other processes. In a practical situation there will be a lot of aspects that have some importance for the production. We will ignore many of these aspects, because they would complicate the problem considerably, without being an essential element for the control rules for production planning and for the lead times.

We make the following simplifying assumptions. Raw material is always available, machines have no breakdowns and their speed is constant. The set-up times between orders for the same type will be ignored and the set-up time and the set-up costs between different types are independent of the types. The normally available capacity is fixed and if extra capacity is available, the available amount is unrestricted. We can distribute the clients over several groups, with more or less the same wishes about the delivery times and, except in some special situations, the distribution of the demand of each of these groups is known and stationary.

2.2. Costs as instruments for decision making

Our objective is to find good control rules for the production planning and for establishing the lead times. The results of these rules must satisfy both the firm and the client. The performance of a control rule, which can also be seen as a degree of satisfaction, can be expressed in financial terms. The financial aspects may consist of both costs and revenues. The costs can be divided into holding costs, production costs, set-up costs, costs of disservice and costs due to changes in capacity. Some of these costs may be real costs, but they can also be management variables which are used to indicate a level of satisfaction. According to this definition the best control rule is the control rule with the highest 'profit' for the firm, because this rule offers the highest level of satisfaction. First we will describe the different costs, using some of the descriptions given by Silver and Peterson (1985).

The holding costs, also described as the costs of carrying items in inventory, include the opportunity cost of the money invested, the expenses incurred in running a warehouse, the costs of special storage requirements and material handling costs, deterioration of stock, obsolescence, insurance and taxes. According to Silver and Peterson the largest portions of the costs are usually made up by the opportunity costs of the capital tied up in the stock, that could otherwise be used elsewhere in the firm and the opportunity costs of warehouse space claimed by inventories. Neither of these costs can be measured exactly. Although estimating the holding costs may be possible, we consider it to be reasonable, as Brown(1967) argued, that the carrying costs are a top management policy variable, which can be changed from time to time to meet changing environmental conditions. In the situation in which have production to order, holding costs only occur if an order has been produced too soon. Then the order has to be stored until the due date, or it can be delivered earlier to the client. Therefore we will not have large holding costs in the situations that we consider. In our models we will assume the holding costs to be constant over time and to be linear to the number of (unit) items that will be stored.

The production costs can be divided into two parts: the fixed costs and the variable costs. The fixed costs are the costs that would also be charged if no production took place, that is the costs of machines, the warehouse and the personnel. The variable costs are the extra costs due to the production. This may include the costs for energy, for raw material and costs for the deterioration of the machines. As with the holding costs, the calculation of the real production costs will also be difficult. Since most of the production costs will be either constant or linear and therefore not influenced by the production rule or the control rule for lead times, we will ignore the production costs in our models and assume that they are included in the revenues for the orders.

The set-up costs will be considered apart from the production costs. They also contain several elements, such as the costs of the administrative work, the wages of skilled people who have to adjust the machine for the production of another type of product, higher scrap costs and costs due to the lower effectiveness of the machine just after the set-up. Of course no production takes place during the set-up. Therefore the loss of revenues during that period can also be considered as set-up costs. In our models we will assume that the set-up costs and the set-up time do not depend on the product types that are involved. Both the costs and the average set-up time will be constant over time.

If a product is not finished by the due date, costs for disservice are incurred. We will refer to these costs as penalty-costs. In these costs several expenses may be included: costs of administrative work, a price reduction for the client, direct loss of revenues through lost sales, buying the product somewhere else or substitution of a less profitable item. Furthermore, costs are involved because of the goodwill that is lost as a result of the inability to serve. The disservice can affect the future demand of the customer and also the future demand of his colleagues. Because of the rather vague notion of the real cost, the penalty-costs are often considered to be a management decision variable, by means of which the level of disservice can be influenced. In our models we will assume that the demand is always backordered and that the penalty-costs will depend on the priority of an order, which can be type dependent. Furthermore, the costs will be linear to the number of items and to the lateness, the amount of time between the due date and the delivery date.

If the capacity is fixed at a particular level, the wages for the labour forces and the costs for using the shop floor are fixed. These costs are usually considered to belong to the fixed part of the production costs. If the capacity is not used to its full extent, the variable costs of production will be smaller. However, if there is not enough fixed capacity in a certain period, it is sometimes possible to extend this capacity. By doing this, penalty-costs can be avoided, but of course there are also hiring costs related to the temporary enlargement of the capacity. Labour forces have to be paid extra for working overtime, paper work has to be done, meals should be paid for and of course the variable costs of production have to be paid for. In some of our models the level of capacity is fixed and in other models an unlimited amount of extra capacity is available at a fixed price per time unit.

The structure of the revenues is much more simple than the cost structure. In most cases the revenues will be the price that customers have to pay for their orders. This price will depend on the amount ordered and also on the priority class of the order. This priority class sets requirements for the length and the accuracy of the lead times and therefore also influences the penalty-costs. The demand itself may also depend on the lead times or on delivery times. In our models we will assume that the revenues depend on the length and the accuracy of the lead times, whereas the revenues can be different for different priority classes.

The ultimate purpose of the control rules for generating due dates, order acceptance and production planning will be the maximising of the 'profit', that is the sum of the revenues minus the sum of the costs. Maximising the profit can also be considered as giving the shortest possible lead times, with maximum accuracy avoiding too many set-ups with the lowest possible level of capacity and a minimum of overtime. The control rules that will be used for achieving this goal, will now be described more closely.

2.3. Control rules for production planning

Numerous rules for production planning have been studied in the literature. We may distinguish continuous review and periodic review rules, rules for capacitated and uncapacitated situations and for deterministic and stochastic demand. These control rules have been developed to serve different purposes: some rules are intended to smooth the production level, others are developed to minimise the costs or to minimise the lead times. Now which of the well-known rules could be of any help to us in the situation with production to order?

Most of the rules in periodic review situations are based on a deterministic demand or a demand that is known completely for several periods. We assume the demand to be stochastic and that new orders can be placed in all but the current period. Therefore we need rules that make decisions on a rolling-schedule basis: every period the scheduling for a number of periods is done, but only the schedule for the first period is implemented. A lot of work has been done on this dynamic lot-sizing. The names of

Baker (1977), Carlson and Kropp (1980), Dixon and Silver (1981), Dogramaci et al. (1981) and Maes and Van Wassenhove (1986) should be mentioned in this context. One of the most simple production control rules, the Silver-Meal rule (Silver and Meal (1973)), can be adapted for our situation, by adding costs for backordering and by replacing the unknown future demand by the expected future demand. The same can be done with the well-known Wagner-Whitin rule (Wagner and Whitin (1958)).

Now we will come to the method which will be used most to deal with the problem and which can be used in all situations, continuous review and periodic review, capacitated or with flexible capacity. In this method, we replace the complex state, being the demand for one type, distributed over several periods and possibly over several priority classes, by one aggregated variable: the number of penalty points. The number of penalty points for a certain type indicates the urgency of the production of the type. The number is based on the demand for that type and it can be a function of the number and size of the orders, the residual lead time and the priority of the orders. These penalty points are used to make decisions about the type that has to be produced. The decision rule will, in general, take a very simple form: if we have to choose between different types, we shall choose the type with the highest number of penalty points. For types with the same number of penalty points we will use a fixed sequence, depending on the average demand. Usually, a second rule will be added to this choice-criterion: we will only consider those types for which the number of penalty points is sufficient to justify a set-up. This sufficiency will be expressed in a single value, the penalty-minimum, which can be different for different types.

2.4. Control rules for lead times and order acceptance

Only very simple decision rules will be used for the lead times and the order acceptance. We shall not consider any rules for making agreements about fixed lead times. In the situations in which we have fixed lead times we assume that we have different groups of 'clients' and that every group obtains a different fixed lead time, according to an agreement that was made beforehand.

In the situation in which we have firm-initiated lead times, the due dates will usually be based upon the expected production date, but there can also be a maximum lead time. This maximum lead time offers the client some certainty and its value may be different for different groups of clients. For the acceptance of the promised lead times we will consider two models. In the first model all lead times are accepted by the clients and we shall use the average lead time as an element for the performance of the control rule. In the second model the acceptance of a lead time is a stochastic process in which the probability that a client belonging to a certain priority group accepts a certain lead time for his order is given and depends on the group and on the lead time. In this second model the revenues will depend upon the number of orders, for which the lead times are accepted.

Chapter 3

A FIXED PRODUCTION CYCLE

3.1. Introduction

In a situation that is ideal considering the reduction of uncertainty in production, agreements are made with all customers about all deliveries over a longer period, for instance a year. The amount and the delivery date have been settled for all orders and this has been done with respect to a production plan in which the production has been organised in such a way that the number of set-ups will be limited and the available capacity will be sufficient. This ideal situation is of course very difficult to realise, since most of the customers do not have sufficient insight into their future demands and they want to be able to cope with unforeseen situations. The supplying firm may have to face unforeseen situations too, such as strikes, long-lasting machine repairs and unexpected new clients. Therefore the ideal situation, with no uncertainty for a long period, can only be seen as a goal and not as a real-life situation.

For both the supplying firm and its customers, the next best thing to this ideal situation, considering the reduction of uncertainty in the production, is perhaps a situation with a fixed production cycle. In this text, the following definition of a fixed production cycle will be used. *A fixed production cycle is a cycle such that the sequence in which the types will be produced as well as the available capacity for the production of a type is fixed. The production cycle is repeated over and over again.* A production cycle may contain every type exactly one time, but it is also possible that some of the types are produced several times during one cycle. Every occurrence of a type in a cycle will be called a production opportunity for the type. A production opportunity is preceded by a set-up for the type. The available capacity for the production of a type, the so-called length of the production interval, is the same every cycle, but may be different for different types. A similar cycle for the deterministic case with production to stock has been considered by Silver and Peterson (1985, pp. 433-435).

In order to obtain good results with a fixed production cycle, it is necessary that the demand is stationary, not too irregular or lumpy and that a lot of customers are able to give good estimates of their future orders, thus enabling the firm to determine a production cycle based on the expected demand. A smooth demand, which does not deviate much from its forecast, ensures the accuracy of the allocations for capacity and raw material and the avoidance of extra costs for shortages, working overtime and storage. The effect of this smooth demand is also that the customers know when they have to order and when the orders will be delivered. The accuracy of the demand forecast depends on the percentage of the capacity for which contracts are concluded and on the disturbance by irregular clients.

Using a fixed production cycle may have advantages for both the clients and the supplying firm. Clients have more certainty about when they have to order and when an accepted order will be completed. The delivery date may be during the first production opportunity or during one of the following production opportunities. The firm knows when the product types will be produced and this may be an advantage for controlling the inventory level of the raw material and for planning repairs and maintenance. Another positive effect is that the production planning will be much easier and therefore can be done at lower costs.

Of course, there are also disadvantages in the use of a fixed production cycle, especially if there is a widely varying demand. A widely varying demand leads to an ineffective use of the machines: very often there will be no more orders for the type for which the capacity reservation was made, whereas at the same time a lot of orders for other types of products may be waiting. This leads to fluctuating delivery times and uncertainty for the clients, because some of the orders will only be delivered after several cycles. Therefore, they do not know when they have to order. Another disadvantage of the fixed production cycle is the impossibility to deal with priority orders in the correct way. If a very urgent order is placed, it will not be produced until the next production opportunity and this production opportunity is independent of the demand.

In later chapters the use of a fixed production cycle will be compared with non-cyclic production rules. To make a reasonable comparison possible, the optimal fixed production cycle has to be determined. Therefore the optimal values for the available capacity for all product types have to be determined, with respect to profit maximisation or cost minimisation. An important element in this optimisation is that the use of the machines can be improved by making the fixed cycle longer, but this has a negative effect upon the delivery times. In the following subsections the delivery times and the costs of the fixed cycle production rule will be analysed for three different service disciplines with different demand and service time distributions.

3.2. Analysis of fixed production cycles

In the analysis of the fixed production cycles we will study the situation with one machine, on which M different types of products will be manufactured. The machine is assumed to be perfect, that is to have no breakdowns, and the set-up cost and the set-up time are independent of both the previous type which is produced and the next type that will be produced. The set-up time will be deterministic. Due to the production rule it is known exactly when a particular type will be produced. This knowledge can be used for determining exact due dates. Therefore the arriving orders will be scheduled according to the first-come-first-served rule (FCFS) and the due date will be based on this schedule. Since this schedule will not be changed, there will be no due date deviation and therefore no holding costs or penalty costs. The analysis of the production rule will be quite difficult if the last order at the end of a production interval is always finished, the so-called non-pre-emptive service discipline. Because of these difficulties we will assume that the work on the last order continues in the next production interval of the same type, the so-called pre-emptive service discipline.

A lot of research has been done on problems with cyclic production rules. The difficulty of the analysis depends on the service discipline. Problems where service continues until a queue is empty, the so-called exhaustive service, have been treated by Eisenberg (1970), Swartz (1980), Watson (1984) and others. These authors have found general results on average delivery times and on the average amount of work in process. Problems with other service disciplines, for instance where only one customer of each type is served per cycle (ordinary service), seem to present considerable analytical problems. Some special cases, with only one or two different types, have been analysed exactly, for instance by Cohen and Boxma (1983). In the situation we shall discuss here, we assume that for every type of product a fixed time slot is available during one production cycle. Therefore the analysis of our situation will be different from the analysis presented in the papers that deal with the problems with either exhaustive service or ordinary cyclic service.

Suppose that during one production cycle work can be done on type s_1 during a time interval with length c_1, then there is a set-up, after which work can be done on type s_2 during a time interval with length c_2 and so on until finally after working on type s_m and a set-up, the sequence is repeated and the work on type s_1 can start again. Of course every type of product should occur at least once in this sequence, but a sequence may also contain some types more than once, interrupted by other types. In this way a fixed production cycle can be described by two sets, both containing m elements, with $m \geq M$. The first set is the sequence set $S=\{s_1,s_2,...,s_m\}$ describing the sequence in which the types are produced during one production cycle . The second set is the capacity set $C=\{c_1,c_2,...,c_m\}$ in which the available capacity for the production of the corresponding element in the sequence set is described.

These two sets S and C are not the only important characteristics of the fixed cycle. The service discipline is also very important. During the production interval of a type we can only work on orders of that type. The orders of that type will be produced according to the FCFS-rule. Because of the exact due dates this rule is the same rule as the earliest-due-date rule (EDD). If there are no (more) orders of a type we have to wait for new orders of this type or for the end of the production interval. This way of service, in which we stop working on orders from a certain type at the end of a production interval, is usually called gating service. We will consider two different ways of gating service.

In the first situation with gating service we can work on all orders of a type during the production interval of this particular type. In the second situation we can work only on those orders for this type that arrived before the decision moment. The decision moment is the moment just before starting the set-up to that particular type. The first situation will be called extended gating service and the second situation normal gating service. Extended gating service yields shorter delivery times, because orders that arrive during the production interval may be served immediately and therefore this kind of service will be preferred. In many situations however, extended gating service is impossible or very expensive. Possible reasons for this may be that the sequence in which the orders of the same type will be produced should be known at the decision moment, possibly because there are small set-ups between these orders or for quality reasons, or because decisions on raw material needed for the production of the orders have to be made at the decision moment.

If we consider extended gating service, a set-up will always be performed before the beginning of a new production interval even if there are no orders, because it is possible that new orders will arrive during the production interval. In the situation of normal gating service we have no set-up if there are no orders, because here only those orders will be produced, that are known at the decision moment. We shall consider two different production models with normal gating service, one in which working overtime is possible and one without this possibility. Working overtime means here that work can be done on a type after the end of the production interval if not all orders are finished that should be finished by the end of the interval. This work will be done outside the normal working hours or may even be performed by another firm. In any case, there are no consequences for the available capacity and for the start of the production interval for the following type that will be produced. Next we will describe the analysis of the different service disciplines.

3.2.1. Extended gating cyclic service model

In this subsection we shall consider the situation in which we have extended gating service with no possibilities for overtime. During the production interval of a type we can work on any order of this type, but not on orders of another type. If somewhere during the production interval there are no (more) orders of the type, we must wait until the interval is finished or until new orders arrive. Jobs that are not finished by the end of the production interval have to wait until the next production opportunity. In the next production interval the work on these orders will be continued, so no extra work outside the normal working hours will be done. The set-up will be done just before the production interval. We are especially interested in the average delivery times for the different types. Therefore we must make some assumptions about the demand and service time distribution and about the sequence set S, which may enable us to calculate the delivery times.

For every type of product the orders are supposed to arrive according to a Poisson process with parameter λ_i, $i=1,..,M$ and the service time for an order of type i is, for simplicity, supposed to be exponentially distributed with parameter μ_i, $i=1,...,M$. Also for simplicity it is assumed that the sequence set is given by $S=\{1,2,..,M\}$, implying that every type of product is produced exactly once during a production cycle. The time needed for one set-up is a fixed constant s and the capacity set is given by $C=\{c_1,...,c_M\}$. One production cycle consists of M set-ups and M production intervals, one for every type. Thus the length of a production cycle is given by

$$T=sM+\sum_{i=1}^{M} c_i \tag{3.2.1}$$

We want to avoid that the queue length for some of the types becomes infinitely long. Therefore it is necessary that for every type the available capacity per cycle is large enough to produce the average demand during one cycle. This can be described by the following restriction for the capacity set.

$$\mu_i c_i > \lambda_i T \qquad i=1,2,..,M \qquad (3.2.2)$$

For this situation we first want to determine the average delivery time for every type of product for a given set C for which (3.2.2) holds. Then an approximation method will be presented by which we can estimate the average delivery times for a given set C. This approximation method can help us to determine an optimal set C, for which we have the highest profits or the lowest average costs. One element in the costs will usually be a weighted combination of average delivery times.

Suppose the process started at minus infinity, so we are in a stationary situation. Let us consider one type of product during one arbitrary cycle starting at 0. Omitting the subscripts, we denote the time we are not working on the type by X, $X=T-C$. This will be the first part of the cycle; the second part, the interval with length C between X and T, is the time we can work on the type. The set-up time is included in X. This is illustrated in Figure 3.1.

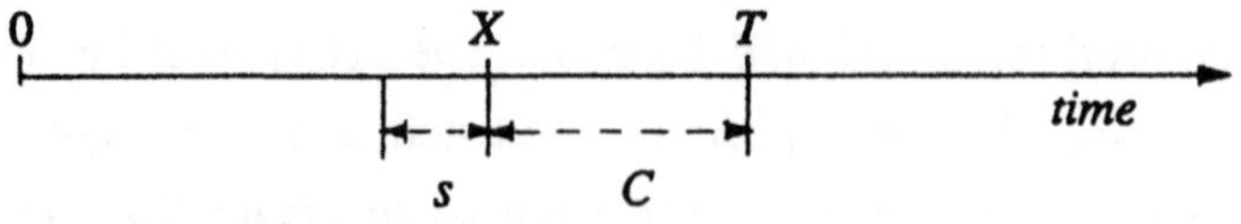

Figure 3.1 The production cycle for a single type

Let L be the average (long run) number of orders in the queue. For an order arriving in the arbitrary interval $[0,T)$, the delivery time can consist of three parts:

(1) if it arrives at time $y \in [0,X)$ it has to wait, $X-y$ until the first production opportunity comes;

(2) if it finds k other orders of the same type in the queue upon arrival, $k+1$ services have to be completed before the order is finished;

(3) if the order is finished in the l-th production interval after its arrival, then $l-1$ times X should be added to the sum of (1) and (2).

Although these three parts are not independent, we may consider them separately in order to determine the expected delivery time of an arbitrary order arriving in the $[0,T)$-interval.

Since the orders arrive according to a Poisson process, we can use the following property, stated for instance by Ross (1972). This property says: suppose that we know that n events, $n \geq 1$ of a Poisson process have occurred by during the interval $[0,T)$. Then the set of n arrival times has the same distribution as a set of n random variables which are independent and uniformly distributed on the interval $[0,T)$. Therefore, the contribution of (1) to the expected delivery time of an arbitrary order, written as $\mathbb{E}(1)$, is given by:

$$\mathbb{E}(1) = \frac{1}{T} \int_0^X (X-y)\, dy = \frac{X^2}{2T} \tag{3.2.3}$$

The contribution of (2) to the expected delivery time, written as $\mathbb{E}(2)$, can easily be calculated using Wald's equation and thus gives:

$$\mathbb{E}(2) = \frac{L+1}{\mu}$$

The contribution of (3) to the expected delivery time, written as $\mathbb{E}(3)$, can be determined in the following way. If the expected number of unfinished orders at the end of the production interval, is given by E, then the average contribution of element (3) to the sum of the delivery times of all previous orders is given by XE. But since this contribution returns every cycle, the expected contribution of element (3) to the sum of the delivery times of all orders arrived in $[0,T)$ is given by XE. The expected number of orders that arrive during a production cycle is given by λT. Using Little's formula, the contribution of (3) to the expected delivery time of an arbitrary order will be given by:

$$\mathbb{E}(3)=\frac{XE}{\lambda T}$$

Using the results for $\mathbb{E}(1)$, $\mathbb{E}(2)$ and $\mathbb{E}(3)$, the expected delivery time for an arbitrary order, denoted by D, is given by:

$$D=\frac{(L+1)}{\mu}+\frac{X^2}{2T}+\frac{XE}{\lambda T}, \tag{3.2.4}$$

Since we have Poisson arrivals, we can use Little's formula for the average number of orders in the queue:

$$L=\lambda D \tag{3.2.5}$$

and the only remaining problem is the calculation of E.

In order to calculate E, we introduce P_k, the equilibrium probability that the number of unfinished orders for the type equals k at the end of an arbitrary production interval. If these probabilities are known, we can determine E by:

$$E=\sum_{k=1}^{\infty} k P_k \tag{3.2.6}$$

Suppose there are i unfinished orders at the end of an arbitrary production interval and suppose that j new orders arrive before the beginning of the next production interval. We will write $p^C_{i+j,k}$ for the probability that, starting with the $i+j$ orders at the beginning of the next production interval, there will be k unfinished orders at the end of this next production interval. Here the superscript C comes from the length of the production interval. Using this probability, P_k should satisfy:

$$P_k=\sum_{i=0}^{\infty} P_i \sum_{j=0}^{\infty} \frac{(\lambda X)^j}{j!} e^{(-\lambda X)} p^C_{i+j,k} \tag{3.2.7}$$

The probability $p^C_{i+j,k}$ is by its definition the same probability as the probability that the number of orders in the queue equals k at time C, given that it equals $i+j$ at time 0, $\mathbf{P}(X_C=k \mid X_0=i+j)$, in an M|M|1 queue with $0<\rho=\frac{\lambda}{\mu}<1$. A lot of expressions have been derived for this probability, for instance by Asmussen (1987) and Prabhu (1965). The most common expression is given by

$$p^C_{i+j,k}=l_{k-i-j}+\rho^{-i-j-1} l_{i+j+k+1}+(1-\rho)\rho^k \sum_{n=-\infty}^{-i-j-k-2} l_n \tag{3.2.8}$$

with

$$l_n = e^{(-(\lambda+\mu)C)}\rho^{n/2} I_n(2\delta C) \tag{3.2.9}$$

where $I_n(x)$ is the modified Bessel function of integer order n, defined by

$$I_n(x) = \sum_{k=0}^{\infty} \frac{(x/2)^{2k+n}}{k!(k+n)!}, \quad I_{-n}(x) = I_n(x), \qquad n \in N \tag{3.2.10}$$

and $\delta = \sqrt{\lambda\mu}$. From these formulae the steady-state probabilities P_k can be calculated as accurately as we want.

Due to the complicated form of E, it will be very difficult to minimise the costs for a given situation and to find the optimal capacity set **C**. Of course some capacity sets can be tried, but it would be nice if we had some reasonable set to start with. Therefore we have to approximate E by something that is much easier to calculate. There are several options for this approximation. For a short production cycle the average will tend to be like the average found in the so-called shadow approximation. Let ρ^* be defined as

$$\rho^* = \frac{\lambda T}{\mu C} \tag{3.2.11}$$

Then the average number of orders in the shadow approximation, E_s, is given by

$$E_s = \frac{\rho^*}{1-\rho^*}$$

At the end of the production interval, it is reasonable to expect that E will be smaller than the value E_s suggested by the shadow approximation, but due to the delayed service E will always be larger than the average number of orders, E_M in the M|M|1-queue with $\rho = \frac{\lambda}{\mu}$, which is given by

$$E_M = \frac{\rho}{1-\rho}$$

From several examples we learned that the value of E is usually situated between the two averages suggested by these approaches. Therefore we choose to replace E by $\hat{E}$, a combination of the average number of orders E_s and E_M of the following form

$$\hat{E} = \hat{\rho} E_s + (1-\hat{\rho}) E_M \tag{3.2.12}$$

For these examples we tried several logical choices for $\hat{\rho}$, all containing in some simple form the elements λ, μ, T , X and C. The best choice found for $\hat{\rho}$ was: the expected length of the busy period divided by the length of the production interval, i.e.:

$$\hat{\rho} = \frac{\lambda X}{(\mu - \lambda)C} \tag{3.2.13}$$

Using (3.2.12) and (3.2.4), we now have a simple expression for an approximation of the average delivery time, which can be used to determine a reasonable capacity set if we want to minimise the average costs or maximise the profit. Comparing the average delivery times with the estimated delivery times using $\hat{E}$ and the shadow approximation for several examples, results in the following table.

Average delivery times					
Example	ρ	ρ^*	Exact	$\hat{E}$ approx.	shadow
1	.250	.891	4.006	4.373	2.736
2	.333	.888	4.875	5.085	3.973
3	.250	.830	5.746	5.815	4.888
4	.333	.927	4.173	4.493	3.184
5	.250	.906	4.327	4.671	3.195
6	.167	.865	4.425	4.792	3.207
7	.083	.766	4.465	4.772	3.266
8	.150	.429	0.391	0.400	0.250

Table 3.1 Comparison between exact and approximated delivery times

The difference between the approximated time using (3.2.12) and the exact average delivery time is in all examples between 2 and 9 percent, whereas the shadow-approximation leads to a difference between 15 and 36 percent, always in the opposite direction. From this random set of examples, we can already conclude that by using $\hat{E}$ as an approximation, we can estimate the delivery times quite well. Another important aspect is that the value of several cost functions containing the weighted sum of the delivery times of the different types could be estimated very well by the approximated delivery times using $\hat{E}$. Although the values of the approximated delivery times are not exactly correct, the capacity sets C, for which the approximated cost functions are minimal, do yield almost minimal costs, due to smoothness of the cost function in the direction in which the capacity set is different from the optimal capacity.

3.2.2. Normal gating cyclic service with strict capacity

This model shows much resemblance with the extended gating service model and we will use the same demand process, the same service time distribution and the same sets S and C. There are two main differences between both models. The set-up for a type will only be performed if there are orders for that type at the decision moment, that is the moment at which the set-up would start. During the production interval only the orders known at that decision moment will be produced. If there are no orders at the decision moment, we will wait during the time needed for one set-up, which is the fixed constant s, and the reserved production interval for the type with no demand, before considering a set-up for the next type at the following decision moment. The main reasons for this kind of service is that we want to consider a fixed cycle that is really fixed and we do not want penalty costs and holding costs. Thus the length of the production cycle, T, will still be given by (3.2.1) and inequality (3.2.2) should still hold for the capacity set.

As well as in the extended gating service model, the delivery time consists of three parts, but these parts are slightly different from the parts in the extended gating service model. Again we shall consider one type of product during an arbitrary production cycle starting at 0. This time the cycle starts with an interval of length s, intended for a set-up, but if there are no orders for this type it can also be waiting time. The second part of the cycle is an interval with length C, during which we can work on orders of the type. The last part of the cycle, with length $X-s$, with $X=T-C$, is meant for the set-ups and production of the other types. Now the three parts of the delivery time of an order arriving in this arbitrary interval $[0,T)$ are:

(1) an order arriving in this interval has to wait until the first production opportunity comes at $T+s$.

(2) the orders whose service times contribute to the delivery time of the arriving order at time $y \in [0,T)$ are the orders that arrived before 0, which are not finished at the end of the production interval, that is at $C+s$, and the orders that arrived between 0 and y.

(3) if the order is finished in the production interval $[lT+s, lT+C+s)$, then $l-1$ times X should be added to the sum of (1) and (2).

To determine the expected delivery time of an arbitrary order arriving in the $[0,T)$ interval, we will consider the three parts separately, as we did in the previous model. Then the contribution of (1) to the expected delivery time, written as $\mathbb{E}(1)$, is given by:

$$\mathbb{E}(1) = \frac{1}{T}\int_0^T (T+s-y)\,dy = \frac{T}{2}+s$$

To describe the contribution of (2) to the expected delivery time, written as $\mathbb{E}(2)$, we will define E, now denoting the average number of unfinished orders at the end of the production interval, in this case at $C+s$, that arrived before the decision moment preceding this interval, in this case at time 0. In this normal gating service model, orders arriving during the production interval or during the set-up, that is between 0 and $C+s$, do not have to wait for all orders that are present on arriving, but only for those orders that have not been produced by the end of the interval, at $C+s$. Now the expected sum of those service times which have an influence upon the delivery time is given by:

$$\mathbb{E}(2) = \frac{E+1}{\mu} + \frac{\lambda}{\mu T}\int_0^T y\,dy = \frac{E+1+0.5\lambda T}{\mu}$$

Using the same reasoning as in the extended gating service model, the contribution of (3) to the expected delivery time, written as $\mathbb{E}(3)$, can be described by:

$$\mathbb{E}(3) = \frac{XE}{\lambda T}$$

Using the results for $\mathbb{E}(1)$, $\mathbb{E}(2)$ and $\mathbb{E}(3)$, the expected delivery time for an arbitrary order, denoted by D, is given by:

$$D = \frac{T}{2} + s + \frac{(E+0.5\lambda T+1)}{\mu} + \frac{XE}{\lambda T}, \qquad (3.2.14)$$

The remaining problem is again the calculation of E.

To determine E, we introduce again P_k, now defined as the steady state probability that the number of orders present at the decision moment, but not produced at the end of the production interval, equals k. If these probabilities are known, we can determine E by (3.2.6). During one production cycle the number of unfinished orders increases due to new arrivals during an interval with length T and it decreases due to services during an interval with length C. These two processes can be considered apart, with only one exception: the resulting number of unfinished orders cannot be negative. Now if we write l_j for the probability that during one production cycle the number of orders will increase by j, under the assumption that the machine produces during the complete production interval. Then P_k should satisfy:

$$P_k = \sum_{i=0}^{\infty} P_i \, l_{k-i}, \qquad k=1,2,\ldots \qquad (3.2.15)$$

and

$$P_0 = \sum_{i=0}^{\infty} P_i \sum_{j=-\infty}^{-i} l_j \qquad (3.2.16)$$

The expression for l_j, which can be seen at the probability that the difference between two Poisson processes takes the value j, shows much resemblance with (3.2.9) and is given by Asmussen (1987):

$$l_j = e^{(-\lambda T - \mu C)} \rho^{j/2} I_j(2\delta), \qquad (3.2.17)$$

with

$$\rho = \frac{\lambda T}{\mu C} \qquad (3.2.18)$$

and $I_j(x)$ is the modified Bessel function of integer order j, defined by (3.2.10), now with

$$\delta = \sqrt{\lambda T \mu C} \qquad (3.2.19)$$

From these formulae, we can calculate the steady-state probabilities P_k as accurately as necessary and thereby E and the expected delivery time D. The calculation of E is much easier than in the extended gating situation. Therefore we will not give an approximative method for estimating E in this situation, although we expect that the method for estimating E will be quite similar to the method described in 3.2.2.

3.2.3. Normal gating service with flexible capacity

In a situation with normal gating service and strict capacity, clients do not quite know when they have to order their products. Sometimes their order will be delivered during the next production cycle, but it may also take several cycles before they will be delivered. If it happens rather frequently that an order will not be delivered during the next production cycle, the clients may become dissatisfied and they may prefer another supplier. The supplying firm can avoid this kind of situation and guarantee, for instance, that all orders will be delivered no later than after the first complete production interval of their type. This means that all orders for a type which are present at the decision moment will be produced during the production interval, or if not all orders are finished by the end of the production interval, extra work will be done for those orders that were already present at the decision moment. This extra work will be done outside the normal working hours, or maybe by another firm. We assume that the available capacity and the start of the production intervals for other types of products are not influenced by this extra work.

In later chapters, we will also consider the possibility of doing extra work in combination with a non-cyclic production rule. In those situations we will assume that the available capacity is the same constant for every production interval. This means that the capacity set C takes the form: $(c,c,..,c)$. Therefore we will use the same form for the capacity set in this situation with normal gating service with flexible capacity. In this situation we also consider different forms for the set S, such that some types occur more than once during a production cycle. The most obvious choice for the set S in the situation with a fixed length for the production interval is the choice in which the frequency in which a type occurs in a production cycle is linear to the demand for that type. We can use this kind of choice if we want to minimise the costs of set-ups, overtime and the delivery times.

The calculation of the amount of extra work and the average delivery times for given sets C and S is not very difficult. An example may clarify this. As an example, we can consider the length of the sum of the production interval and one set-up as one period. We assume that the required capacity for an order is fixed and that in the production interval exactly C orders can be manufactured, independent of the type. Suppose a type occurs three times in the sequence set S, so we may work on this type three times during one production cycle, and the number of periods between the production intervals is n_1, n_2 and n_3. If the probability that the total demand during n periods for the type equals k, denoted by P_{kn}, is given, then we can compute the interesting aspects. The probability that we do not have a set-up after n_i periods is given by P_{0n_i}. Therefore, the expected number of set-ups during one cycle for the type that we consider, U_{type}, is given by

$$U_{type}=3-\sum_{i=1}^{3}P_{0n_i}$$

Assuming that the delivery time of an order is expressed in an integer number of periods, so it may take the values $1,2,..,max(n_1,n_2,n_3)$, the average delivery time for this type, D_{type} expressed in periods, is given by

$$D_{type} = \frac{\sum_{i=1}^{3} \sum_{j=1}^{n_i} j}{\sum_{i=1}^{3} n_i} = 0.5 + \frac{\sum_{i=1}^{3} n_i^2}{2\sum_{i=1}^{3} n_i}$$

and the average amount of extra work per production cycle, w_{type} expressed in number of orders, is given by

$$w_{type} = \sum_{i=1}^{3} \sum_{k=c+1}^{\infty} (k-c) P_{kn_i}$$

Of course there may be situations which are more complicated than this example, but as long as it is possible to produce all the demand that is present at the beginning of the production interval, if necessary by doing extra work, the analysis will only be as difficult as in this example. The situations with the fixed production cycle we have considered in this chapter are on the one hand illustrations of cyclic production rules, which seem to us both useful and interesting to analyse, because cyclic rules are often used in practical situations and on the other hand they can serve very well as some kind of measure for the comparison of non-cyclic production rules in the following chapters.

Chapter 4

NON CYCLIC METHODS FOR ONE PRODUCT TYPE

4.1. Introduction

In certain cases the capacity restrictions are not very important, for instance if the products have to be put into a chemical bath and the size of the bath is much larger than the average batch that is produced. It is also possible that the extra costs for working overtime or extra capacity are very small compared to the set-up costs. Also in practical situations in which there are capacity restrictions it can be wise to at first ignore these restrictions in order to get some insight into the other problems. We have assumed that raw material is always available. Therefore, the only way in which the production planning of different product types is related, is through the capacity restrictions. If these restrictions are rather unimportant, it is not necessary to consider all product types together, but the product types can be considered separately in order to analyse the problem. In this chapter, we will consider the production planning for a single type of product, where the problem of our interest is the balance between a limited number of set-ups and short delivery times or a limited number of late orders.

This problem is particularly interesting if the set-up costs are large. Different from the cyclic production rule, a set-up will only be started if there is a sufficient number of jobs to be produced. We will assume that the decisions concerning production are made periodically. The element of periodic review is introduced here because it is helpful in the analysis and because this element is a practical given in a lot of production planning systems. The length of the review period will be fixed. We will assume that the decisions will be made at the end of a review period. At that time all new orders are known. The decision about the production concerns the production in the next period of orders known at the end of the review period. The decisions are not changed during that period, so in the first period we never produce orders that have been arrived during this first period. We have found the same element in the gating service model.

In this chapter we shall consider one rule for determining the lead times and several rules for the production planning of a single type of product on one machine, with large set-up costs. Other costs that occur are holding costs, for orders that are produced too early, and penalty costs for orders that are produced too late. Late orders will be backlogged. In this chapter we will study three different situations. Maybe some elements of these situations will not be very realistic, especially the independent demand, but there are certainly situations where the proposed decision rules can be used. First we will study the situation in which the lead times for the orders are fixed according to a long-term agreement. The distribution of the demand is assumed to be stationary. The firm has to manufacture the orders at minimal costs. We will start with the assumption that the penalty costs are the same for all clients and later on we will give an extension for clients with different penalty costs.

Next, we consider a situation where the lead times are again fixed, but the demand is no longer stationary. The number of clients, and thereby the demand, fluctuates due to the lateness of previous orders. Usually such fluctuations will be included in the variance of the demand and for the late orders penalty costs are assumed. In our model the number of clients is varying according to a stochastic process, which depends upon the number of orders that have been delivered too late.

In the third situation we consider firm-initiated lead times. The length of the lead time for the orders that arrive in a certain period depends on the due dates of the previous orders, which have not been produced yet, and on the number of orders that have been arrived during the period. We assume a simple stochastic model for the acceptance of the proposed lead times by the customers. The probability that a client accepts a lead time will be a decreasing function of the proposed lead time. The decisions about the lead times are based on the probabilities for the acceptance, which are supposed to be known to the firm.

4.2. Fixed lead times

4.2.1. Introduction

In this situation clients place an order and the lead time of the order is fixed according to a long-term agreement. Every order consists of one unit of product, where the size of the units is chosen according to the holding costs and the penalty costs. The lead time of an order can be the next period or one of the $N-1$ later periods. Therefore, in an arbitrary period t, orders may obtain a due date in one of the periods $t+1$, $t+2$,..., $t+N$. The probability distribution of this demand, for each of the N periods, is a given distribution which is independent of t. For our model, we assume that the orders are placed by N groups of clients, where the lead time for orders placed by clients belonging to group i is also equal to i. The demand during period t of the different groups of clients are assumed to be stochastically independent.

A typical element for a situation where we have production to order and fixed lead times, is a gradually decreasing level of knowledge about the required deliveries for future periods. At the end of an arbitrary period, say period t, the required deliveries for period $t+1$ are known exactly, nearly all required deliveries are known for period $t+2$, except the orders that will arrive during period $t+1$. Furthermore there is a decreasing knowledge about the required deliveries for the periods $t+3$ until $t+N$. Clients can still place orders for these periods during the following 2 or more periods. With this knowledge about the required deliveries and with the knowledge of the demand distribution, a decision about the production has to be made.

At the end of period t we can decide to produce a subset of the known orders, to produce all the known orders, or we can decide to delay production. If we produce in period $t+1$, then we have set-up costs s and if we produce orders that have to be delivered during one of the periods $t+2,...,t+N$, we have holding costs h per order per period for the orders that are manufactured too early. If there are required deliveries for period $t+1$ or even earlier, and we decide to delay the production, then we have penalty costs at the end of period $t+1$. The penalty costs at the end of this period will be p units per order for the required deliveries for period $t+1$ or earlier. We have chosen for this linear form because this allows a tremendous simplification of the state space, due to an aggregation of all late orders. The late orders are not lost, but backlogged.

The objective is to minimise the long term average costs per period. First we will model the problem as a Markov Decision Problem. Using the method of successive approximations, the optimal production rule can be determined as well as the average costs of this rule. Then we will describe some heuristic methods and consider their performance in a few examples, as well as the performance of a cyclic service rule. In these examples all heuristics will turn out to perform reasonably well. The main differences between the heuristics are in the complexity of the decision rules and the possibility to compute the average costs.

4.2.2. Description of the model

In order to give a good description of the problem, we shall model it as a Markov Decision Problem (MDP). Markov Decision Processes have been studied initially by Bellmann (1957) and Howard (1960). We will first give a short description of an MDP in general. Suppose a system is observed at discrete points of time. At each time point the system may be in one of a finite number of states, labeled by 1,2,.., M. If, at time t, the system is in state i, one may choose an action a, from a finite space A. This action results in a probability P_{ij}^{a} of finding the system in state j at time t+1. Furthermore costs q_i^a have to be paid when in state i action a is taken.

Returning to our problem, we find that we can only give an adequate description of the state if we use a state vector instead of a single integer value. In this state vector we want to express the required deliveries for the various periods. However, at the end of period t, there is no difference in future costs between required deliveries for period t+1 and required deliveries for earlier periods, such as t, t−1, etc.. Therefore we can limit the order state vector to N components. At the end of an arbitrary period t, the first component denotes the required deliveries for period t+1 and earlier periods, the second component denotes the required deliveries for period t+2, and so on, until the N-th and last component, which denotes the required deliveries for period $t+N$. We will denote the state we observe, $r=(r_1,r_2,...,r_N)$, the order state vector. The set of all the possible states is denoted by R.

The second element of the MDP is the action space $A(r)$. Each state $r \in R$ is associated with a finite non-empty set of actions $A(r)$. Since we have no capacity constraints, we will always produce the demand for an integral number of periods. Therefore, the meaning of action a is that we produce the orders with a residual lead time of a

periods or less, that is $\sum_{i=1}^{a} r_i$ products. Action 0 means that we do not produce. To determine the action space, we choose $A(r)$, $r \in R$, in a reasonable way:

$A(r)=\{0\}$ if r_1 equals 0, since it is clearly optimal to delay production;

$A(r)=\{1,..,N\}$ if $r_1 p > s$, since it is clearly not optimal to delay production, because the penalty costs are larger than the set-up costs;

$A(r)=\{0,..,N\}$ for all other $r \in R$.

The transition probabilities from a state vector r to a state vector z are more difficult to describe than in an MDP with single integer valued states. In every period t, clients belonging to group i can order for period $t+i$ with $i=1,..,N$. The probability that they together order j units during an arbitrary period t for period $t+i$ is known and this probability is denoted by d_{ij}. We assume that the demand for different groups is independent. Let $J \subseteq N^N$ be the set of possible one-period demands $(j_1, j_2, ..., j_N)$. Let $Q_a(r)$ be the state at the end of the next period, just before the new orders are added to the order state, if we have taken action a in state r. This implies that the first a components will be given the value 0 and a left-shift afterwards, for instance $Q_0(r)=(r_1+r_2, r_3, .., r_N, 0)$ and $Q_3(r)=(0,0,r_4,..,r_N,0)$. The probability P^a_{rz}, that on observing state r and choosing action $a \in A(r)$ we enter state z given by

$$z = Q_a(r) + (j_1, j_2, ..., j_N) \tag{4.2.1}$$

is therefore given by $\prod_{i=1}^{N} d_{ij_i}$ for all $(j_1, j_2, .., j_N) \in J$, and 0 elsewhere.

Now we consider the one stage costs of taking action a on observing state r. If we produce, we pay set-up costs and also holding costs if $a \geq 2$. We assume that all holding costs for orders that are produced more than 1 period too early will be paid immediately. This assumption is necessary for the transition probabilities described above. If we do not produce than we have penalty costs. Penalty costs for orders that have been delivered more than 1 period too late will be paid separately in each of the periods. Therefore, the one stage costs of taking action a on observing state r have the following form.

$$q^a_r = s + h \sum_{i=1}^{a-1} i \, r_{i+1} \qquad \textit{if } a>0 \tag{4.2.2}$$

$$q^a_r = p \, r_1 \qquad \textit{if } a=0 \tag{4.2.3}$$

4.2.3. The optimal production policy

By the choice of the action space, all stationary policies have transition probability matrices representing recurrent aperiodic Markov chains. If the number of possible states is limited, i.e. if the one-period demand is bounded, we can determine the optimal production policy. The optimal policy can be determined by a policy iteration method, but we will use the method of successive iteration, as described by Odoni(1969), since this method is faster in our situation. The optimal policy is the policy which achieves the minimum expected costs per transition, which will be denoted by g. Defining the quantity $v_n(r)$ as the total expected costs from the next n transitions if the current state is r and if an optimal policy is followed, the iteration scheme takes the form described in the optimality principle by Bellman (1957):

$$v_{n+1}(r) = \min_{a \in A(r)} [q_r^a + \sum_{z \in R} P_{rz}^a v_n(z)] \; ; \qquad r \in R, n=0,1,.. \qquad (4.2.4)$$

Define $x_n(r)$ by:

$$x_n(r) = v_{n+1}(r) - v_n(r). \qquad (4.2.5)$$

Then according to Odoni for any choice of starting conditions $v_0(r)$:

(i) $v_n(r) = ng + v(r) + O_n(r), r \in R$, where $O_n(r) \to 0$ and where the $v(r)$ satisfy:

$$v(r) = \min_{a \in A(r)} [q_r^a + \sum_{z \in R} P_{rz}^a v(z) - g]. \qquad (4.2.6)$$

This function $v(r)$ will be called the relative costs of a state r.

(ii) $x_n(r) \to g,\ r \in R.$

(iii) $L_2(n) = \max_{r \in R} x_n(r)$ is monotone decreasing in n to g.

(iv) $L_1(n) = \min_{r \in R} x_n(r)$ is monotone increasing in n to g.

(v) Any production rule achieving the minima in (4.2.4) for all $r \in R$ for all $n \geq$ some n_0 achieves the minima in (4.2.6) for all $r \in R$ and has minimal costs per transition.

It is not necessary to consider all possible values for first component of the order state vector r_1. If for some state $r \in R$, action $a>0$ is optimal, then the same action will be optimal for all states $(y, r_2, r_3, .., r_N)$ with $y > r_1$, because r_1 is not involved in the holding costs. Beginning with $v_0(r)=0$ for all $r \in R$, we repeat (4.2.4) until a satisfactory degree of convergence is achieved. It follows that g may be estimated from $L_1(n) \le g \le L_2(n)$ as $g \sim [L_1(n)+L_2(n)]/2$. The range decreases by n and the estimate becomes nearly exact for large n.

4.2.4. Heuristic procedures.

If the number of states we have to consider is very large, it will become very difficult or impossible to determine the optimal policy. Therefore we will also consider three heuristic approaches: a Silver-Meal-like approach, a Wagner-Whitin-like approach and a production rule that we will call the (x,T)–rule. Before a description of the three different rules is given, some variables that will be used in the sequel will be defined.

4.2.4.1. Notation

The following variables are defined:

- X_{tw} is the number of orders arriving during period t for period w, $w=t+1,t+2,\ldots,t+N$, (see Figure 4.1).

	lead time			
arrival date	1	2	3	N
t	$X_{t,t+1}$	$X_{t,t+2}$	$X_{t,t+3}$	$X_{t,t+N}$
$t+1$	$X_{t+1,t+2}$	$X_{t+1,t+3}$	. . . . .	$X_{t+1,t+N+1}$
$t+2$	$X_{t+2,t+3}$	$X_{t+2,t+4}$	. . . . .	$X_{t+2,t+N+2}$

Figure 4.1 The demand for the various periods

- $u_{t,t+i}$ is the expected number of orders during an arbitrary period t for period $t+i$: since we have stationary demand, we will write $u_{t,t+i} = \mathbb{E}\, X_{t,t+i} = \sum_{j>0} j d_{ij} = u_i$.

- b_{ilt} is the probability that during the last i periods before an arbitrary period t, clients order a total number of l orders for this period t:

$$b_{ilt} = \mathbb{P}\{ \sum_{n=t-i}^{t-1} X_{nt} = l \}$$

In Figure 4.1. we can see that the separate parts of the demand for period $t+3$ are placed diagonally. Let J_{il} be the set of all possible one-period demands $(j_1, j_2, ..., j_N)$ for which the sum of the first i components equals l. Due to the stationary demand we can write:

$$b_{ilt} = \sum_{J_{il}} \prod_{k=1}^{i} d_{kj_k} = b_{i,l}.$$

- e_{it} is the expected value of the i-th component of the order state vector at the end of period t, if no required deliveries for period $t+i$ have been produced. The separate parts of the demand are placed diagonally in Figure 4.1.

$$e_{it} = \mathbb{E}(\sum_{n=t+i-N}^{t} X_{n,t+i}) = \sum_{n=t+i-N}^{t} u_{n,t+i} = \sum_{m=i}^{N} u_m.$$

4.2.4.2. The Silver-Meal approach

Applying the idea of the Silver-Meal algorithm (Silver and Meal (1973)), we divide the expected costs of an action by the number of periods involved in this action and we choose that action for which the quotient is minimal. Unlike Silver and Meal we will not stop at the first minimum, but we consider the quotient for all possible actions. The direct costs of action a are given by q_r^a. However, this are not the only costs during the first a periods if $a \geq 2$. If we produce the required deliveries, say in period $t+1$ for two or more periods ($a \geq 2$), it is reasonable to suppose that we will not produce during the next a-1 periods, i.e. in the periods $t+2, t+3, ..., t+a$. This implies that the required deliveries for these periods that arrive during the periods $t+1, t+2, ..., t+a-1$ will be delivered too late, so we will have to pay penalty costs for these orders. Therefore we will not only consider the direct costs q_r^a of action a, but also the indirect costs. It would be difficult to calculate the indirect costs exactly, because in some situations we will not wait a periods before we produce again. However, the indirect costs can be estimated easily by the expected penalty costs during the first a periods, if we only produce in the first period and not in the $a-1$ following periods. Due to this estimation the indirect costs depend on a only and they will be described by the penalty function $p(a)$:

$$p(a) = p\,\mathbb{E} \sum_{k=2}^{a} \sum_{i=2}^{k} \sum_{j=1}^{i-1} X_{ji} = p \sum_{k=2}^{a} \sum_{i=2}^{k} \sum_{j=1}^{i-1} u_{i-j} =$$

$$= p \sum_{i=2}^{a} (a+1-i) \sum_{j=1}^{i-1} u_j \qquad \textit{for } a=2,3,..,N \qquad (4.2.7)$$

If we do not produce, we may have penalty costs, but then these costs are in the direct costs. If action 1 is taken there are no penalty costs until the next decision moment.

$$p(a) = 0 \qquad \textit{for } a=0,1 \qquad (4.2.8)$$

Using this penalty function, q_r^a for the direct costs and $\delta(a)$ as a function that equals 1 for $a=0$ and 0 elsewhere, the production rule takes the following form:
if we observe a state $r \in R$ we take that action a for which

$$\frac{q_r^a + p(a)}{a + \delta(a)} \qquad (4.2.9)$$

is minimal over $a \in A(r)$.

4.2.4.3. The Wagner-Whitin approach

Wagner and Whitin (1958), developed an algorithm for a finite horizon production model where the demand for H periods is known completely and terminates at the horizon H. This algorithm guarantees that the best possible actions are taken during H periods. Compared with this deterministic scheduling problem with known demand and a fixed horizon, we have two extra problems: a part of the required deliveries for the next H periods are still unknown and we have a moving horizon, which implies that it might be profitable to produce orders with a due date after the horizon during one of the first H periods. Baker (1977), Carlson and Kropp (1980) and van Nunen and Wessels (1978) have studied the implementation of finite horizon models, such as the Wagner-Whitin algorithm, in a dynamic situation. In their methods, they solve a multi-period problem and implement the first period decision. This process is repeated every period, thus creating a so-called rolling schedule.

Using the Wagner-Whitin algorithm on a rolling schedule base will generally not lead to an optimal schedule. The interesting result however is that the longest possible forecast horizon is not necessarily the best (Baker (1977)). Carlson and Kropp confirm this result, although many of their examples point towards a conclusion that more information leads to better results. Although in their problem description the gradually decreasing level of knowledge of the demand is not present, it seems likely that some characteristic elements of their results will also appear in our problem. Therefore we will consider different values for the horizon H. Another reason for this is that by choosing a horizon H which is smaller than N, the algorithm will be faster. In our version of the Wagner-Whitin approach, we shall use a branch-and-bound method and consider all possible action sequences $a_1, a_2, .., a_H$ during the first H periods ($H \leq N$). We shall choose the first action of the action sequence with minimal projected costs to be taken this period.

The two extra problems, the partly unknown demand and the moving horizon, will be approached in the following way. In our algorithm we do not consider the demand distribution, but we will replace the unknown future demands by their expected value, still assuming that we can not produced expected orders before their arrival date. The second problem is known as the effect of terminal conditions in the rolling schedule. Baker (1981) studies this problem in a special quadratic production-inventory model. In his solution the terminal conditions are based on the profile of states that occurs in a deterministic finite horizon model. Implemented in a situation with uncertain demand, this solution achieves a near-optimal performance. Although we do not have this quadratic production-inventory model, we will also consider terminal conditions. Therefore we assume some simple production rule to be used after the horizon to measure the effect of an action sequence on later periods.

The production rule we will assume to be used after the horizon is a cyclic strategy which will be denoted by π. This strategy is the equivalent of the Period Order Quantity lot-sizing technique proposed by Gorham (1968) in which the optimal production schedule in the case of constant demand requires a set-up every T periods. We also assume a set-up every T periods for our cyclic production rule π and of course, in periods in which we produce the required deliveries for the first T periods will be produced. The expected costs during one cycle are the costs of one set-up, holding costs for orders that are produced too early and penalty costs which are given by the penalty function in (4.2.7) and (4.2.8): $p(T)$. Therefore, the average costs per period for this rule, written as g_π, are given by:

$$g_\pi = \frac{s + h \cdot \sum_{i=1}^{T-1} i e_{i+1} + p(T)}{T}, \tag{4.2.10}$$

where T is chosen over the values $1,2,...,N$ in such a way that it yields the minimum for g_π.

In our Wagner-Whitin approach we want to find the action sequence $(a_1,a_2,...,a_H)$ with minimal costs. The costs of an action sequence consist of three parts: the direct costs in the first period, the expected costs in periods 2,3,..,H and the indirect costs: the effect of the action sequence on the costs in later periods. The effect of the production of orders with a due date after the horizon will be measured by determining the marginal costs for the expected order state at the end of the H-th period. Therefore we introduce a salvage function $L(.)$, which can be compared with the function $v_n(r)$ in the optimal policy (cf. (4.2.4)).

Let $\mathbf{A}_H$ be the set of all the possible action sequences with H elements. Let z_i be the projected order state vector during the i-th period for an action sequence $\mathbf{A} \in \mathbf{A}_H$. Hence:

-1 $z_1 = r = (r_1,r_2,...,r_N)$ is the order state vector in the first period.

-2 $z_i = Q_{a_{i-1}}(z_{i-1}) + (u_1,...,u_N)$ i=2,..,H, is the projected order state vector during the i-th period (cf. (4.2.1)).

Now the production rule takes the following form: given a state $r \in R$ determine the action sequence $\mathbf{A} \in \mathbf{A}_H$ for which:

$$\sum_{i=1}^{H} q_{z_i}^{a_i} + L(\mathbf{A}) \tag{4.2.11}$$

is minimal over the elements of $\mathbf{A}_H$. The first action of the optimal action sequence is taken for the first period.

In this paragraph we will consider the indirect costs. Let $\mathbf{A} \in \mathbf{A}_H$ be a given action sequence. Then we denote by $l(\mathbf{A}) = \max\{i \mid a_i > 0\}$ the last period in which we produced and by $j(\mathbf{A}) = \max\{i + a_i \mid i = 1,..,H\}$ the last period for which some orders have been produced, or H. We will define a salvage function $L(\mathbf{A})$, which depends on $\mathbf{A}$ only by the values $l(\mathbf{A})$ and $j(\mathbf{A})$. If no orders have been produced with a due date after the horizon, that is if $j(\mathbf{A}) \leq H$, we assume that we will produce the first period after the horizon. For an action sequence $\mathbf{A}$ for which $j(\mathbf{A}) \leq H$ the value of the salvage function $L(\mathbf{A})$ is chosen to be 0.

If in an action sequence $\mathbf{A}$ some orders have been produced with a due date after the horizon, that is if $j(\mathbf{A}) > H$, we assume that the next production will take place in the $j(\mathbf{A})+1$-th period and not in period $H+1$, the first period after the horizon. The expected costs during the periods $H+1$ until $j(\mathbf{A})$ will consist of penalty costs only and using the penalty cost function defined in (4.2.7) and (4.2.8), the penalty costs are given by:

$$C(\mathbf{A}) = p(j(\mathbf{A}) - l(\mathbf{A}) + 1) - p(H - l(\mathbf{A}) + 1).$$

Comparing this with the cyclic strategy π in which we have costs g_π per period during $j(\mathbf{A}) - H$ periods, yields the value of the penalty function $L(\mathbf{A})$.

$$L(\mathbf{A}) = C(\mathbf{A}) - (j(\mathbf{A}) - H) \cdot g_\pi \qquad \text{if } j(\mathbf{A}) > H \quad (4.2.12)$$

$$L(\mathbf{A}) = 0 \qquad \text{if } j(\mathbf{A}) \leq H \quad (4.2.13)$$

By using a branch-and-bound method it is not necessary to consider all the possible action sequences. The number of action sequences we have to consider will be limited in three ways:

1) during the first period the action should belong to the action space $A(r)$ of the order state vector r.

2) we do not produce during periods for which a part of the required deliveries have already been produced;

3) we remove action sequences in which we produced a smaller amount at higher costs than in one of the others.

According to the first two restrictions, the following should hold for an action sequence $a_1, a_2, .., a_H \in \mathbf{A}_H$:

1) $a_1 \in A(r)$ is the action during the first period;

2) $a_i = 0$ if $\max_{j<i}[j + a_j] > i$, which implies that we do not produce during the i-th period because some of the orders with this due date period have already been produced;

3) $a_i \in \{0, 1, .., N\}$ otherwise.

With these restrictions, we can easily determine the best action sequence $\mathbf{A} \in \mathbf{A}_H$, which gives the minimal projected costs in (4.2.11).

4.2.4.4. The (x,T)-rule

By examing the optimal policy closely in some examples, we observed the following two properties. Firstly, the decision whether a production takes place or not, depends almost exclusively on the number of orders for the first period (r_1) and not on the required deliveries for later periods. Secondly, for a given demand distribution, the number of periods for which the orders are produced is nearly always the same. An example may illustrate these properties.

Consider a situation with a maximum lead time of 4 periods, penalty costs p=3, holding costs $h=1$, set-up costs $s=6.5$ and a binary demand per period: $d_{i1}=1-d_{i0}=0.5$ for i=1,2,3,4. The number of possible order states is limited in this example and therefore we can determine the optimal policy. The resulting optimal policy has the following features:

1) We do never produce if $r_1 \leq 1$.

2) The optimal action if $r_1 \geq 2$ is always $a=2$, produce the required deliveries for the first two periods, except for the order states (2,3,0,0), (2,3,0,1), (2,3,1,0) and (2,3,1,1). In these four states we do not produce. The probability of finding the demand in one of these states is quite small.

Also in other examples, we observed the same two properties. These two elements seem to be very important if we want to look for a production rule which is simple, but close to optimal. Therefore, we shall propose the following approach. If the number of orders required for the first period or one of the earlier periods, r_1 is smaller than a decision variable x, then we do not produce during that period. If r_1 is equal to or larger than x, then we will produce during that period and we produce all orders with a residual lead time of T periods or less, where T is also a decision variable. Decisions on the production are made at the end of *every* period, so production may take place less than T periods after the previous one. This rule will be called the (x,T)–rule.

In the unconstrained capacity situation it is rather easy to determine the average costs for given values of x and T. The optimal values for x and T can be found by computing the average costs per period for several values of x and T. First we will show how this calculation is done. Afterwards two properties will be given, which can help us determine the optimal values of x and T.

In order to compute the average cost per period, $g(x,T)$, for any pair (x,T), we consider the order state during one production cycle. The decision whether to produce or not depends exclusively on r_1. This means that there are regeneration points in the stochastic process that represents the order state. These regeneration points can be found at the end of a period in which the required deliveries for the first T periods have been produced. The probabilities for each of the different order state vectors are the same in every regeneration point. Therefore we can determine the average costs by considering the order state during the periods between two regeneration points.

The state description $r=(r_1,r_2,\ldots,r_N)$ provides all necessary information about the order state, but it is not suited for the analysis of situations in which the number of possible order state vectors is large. Defining a state simply by (r_1) is sufficient for the decision, but it does not provide the necessary information about the transition probabilities. The changes in the r_1 value depend on the time past since the last production. Suppose that we have produced the orders with a due date of $t+3$ or earlier during period t. Then the probabilities for the r_1 value at the end of period t are given by the distribution for the stochastic variable $X_{t,t+1}$. If the r_1 value at the end of period t is smaller than the decision variable x, then the following period the value of two stochastic variables, $X_{t,t+2}$ and $X_{t+1,t+2}$, will be added to the r_1 value. If at the end of period $t+1$ the r_1 value is still smaller than x, then the following period the value of three stochastic variables, $X_{t,t+3}$, $X_{t+1,t+3}$ and $X_{t+2,t+3}$, will be added. If at the end of period $t+2$ or one of the later periods the r_1 value is still smaller than x, then the value of N stochastic variables will be added. The probability that the sum of the k stochastic variables, which value is added to the r_1 value, is equal to l, is given by $b_{k,l}$ which we have defined in Subsection 4.2.4.1. From this discussion we can conclude that a state description *(time since last production* $,r_1)$ provides all necessary information for the analysis of this rule.

If we use this state description, the number of possible states may be infinite, if there is a probability that there are no orders during a period, or if the demand can be infinitely large. However, both the decision and the costs of the decision are the same for $r_1=x$ and for $r_1>x$. Furthermore, both the costs of an action and the transition probabilities for the component r_1 are the same for all periods, except the $T-1$ first periods. Therefore we can limit the state space: for the r_1-component to the values $\{0,1,\ldots,x\}$ and for the time-component to the values $\{1,2,\ldots,T-1\}$ or $\{1\}$ if $T=1$.

In order to determine the average costs per period, we consider the finite state recurrent Markov chain, with the states *(time,r_1)* described above. In this chain we have the following transition probabilities.

$$\boldsymbol{P}\left\{(i,x)\rightarrow(1,k)\right\}=b_{1,k} \qquad i=1,2,..,T\text{-}1;\ k=0,1,..,x\text{-}1$$

$$\boldsymbol{P}\left\{(i,x)\rightarrow(1,x)\right\}=1-\sum_{k=0}^{x-1}b_{1,k} \qquad i=1,2,..,T\text{-}1$$

$$\boldsymbol{P}\left\{(i-1,j)\rightarrow(i,k)\right\}=b_{i,k-j} \qquad i=2,..,T\text{-}1;\ k=j,..,x\text{-}1$$

$$\boldsymbol{P}\left\{(i-1,j)\rightarrow(i,x)\right\}=1-\sum_{k=0}^{x-1-j}b_{i,k} \qquad i=2,..,T\text{-}1;\ j=0,..,x\text{-}1$$

$$\boldsymbol{P}\left\{(T-1,j)\rightarrow(T-1,k)\right\}=b_{N,k-j} \qquad k=j,..,x\text{-}1$$

$$\boldsymbol{P}\left\{(T-1,j)\rightarrow(T-1,x)\right\}=1-\sum_{k=0}^{x-1-j}b_{N,k} \qquad j=0,..,x\text{-}1$$

In Figure 4.2. an illustration of the Markov chain is given.

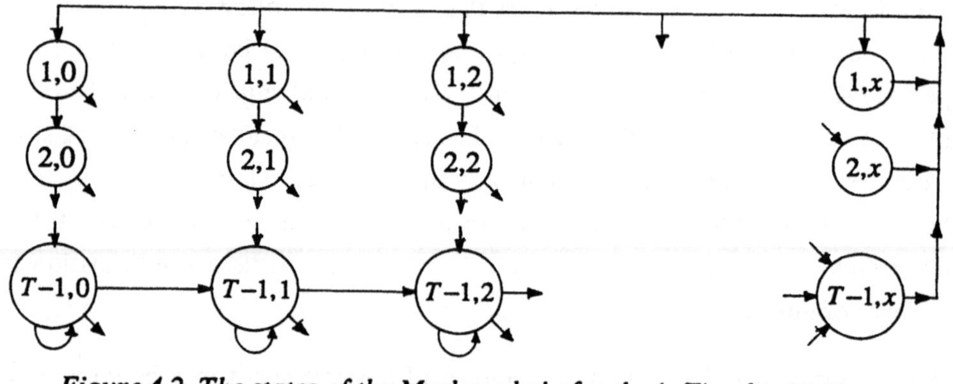

Figure 4.2 The states of the Markov chain for the (x,T)-rule, (T≥2).

In order to determine the average costs per period, we have to determine the expected costs c_{ij} in every possible state (i,j). In most of the states, the only costs are penalty costs and these costs depend on the j-value of the state (i,j):

$$c_{ij}=pj \qquad 0\le j<x,\ 1\le i\le T\text{-}1 \qquad (4.2.14)$$

In the states (i,x), in which we produce, we have set-up costs and usually also holding costs. The holding costs are paid immediately to avoid a complicated state space. The expected holding costs depend on the i-value of the state (i,x), that is on the time past since the last production period, with a maximum of $T-1$. Obviously, at least for $T>2$, if we have just produced in the previous period, the expected number of orders for the future periods will not be as high as they would be after $T-1$ periods, because some of the orders for these periods have already been produced. Thus we can write the expected costs in state (i,x) as:

$$c_{ix} = s + h \sum_{j=1}^{T-1} j e_{j+1} - h \sum_{j=1}^{T-i-1} j\, e_{j+i+1}$$

$$= s + h \sum_{j=1}^{T-1} e_{j+1} \min(i,j) \qquad 1 \le i \le T-1 \qquad (4.2.15)$$

Let q_{ij} be the expected number of visits to the state (i,j) during one production cycle. We can stay in the states $(T-1,0),(T-1,1),\ldots,(T-1,x-1)$ longer than one period. In all other states we can stay only on period. Therefore the expected number of visits to these states is the same as the steady-state probability to visit this state during one production cycle. Due to this definition of q_{ij}, the sum of the expected number of visits to all states, is the expected time between two production periods. Hence:

$$T(x,T) = \sum_{i=1}^{T-1} \sum_{j=0}^{x} q_{ij} \qquad (4.2.16)$$

Because of the special structure of the Markov chain we can determine the values for q_{ij} very easily. In the first period after a production period, we are either in state $(1,x)$ or in one of the states $(1,j)$:

$$q_{1j} = b_{1,j} \qquad 0 \le j < x \qquad (4.2.17)$$

$$q_{1x} = 1 - \sum_{k=0}^{x-1} b_{1,k} \qquad (4.2.18)$$

From state $(i-1,k)$ we move to state (i,j) with probability $b_{i,j-k}$:

$$q_{ij} = \sum_{k=0}^{j} q_{i-1k}\, b_{i,j-k} \qquad 0 \le j < x, 2 \le i \le T-2 \qquad (4.2.19)$$

We move to state (i,x) if the number of orders was less than x in period $i-1$ and at least x in period i:

$$q_{ix} = \sum_{k=0}^{x-1} \left[q_{i-1k} - q_{ik} \right] \qquad 2 \le i \le T-2 \qquad (4.2.20)$$

State $(T-1,j)$ is a state which we can visit during period $T-1$, but also in later periods. We can enter this state both from a state $(T-2,k)$ as well as from a state $(T-1,k)$ and furthermore, we can also stay in the same state with probability $b_{N,0}$, the probability that during the N previous periods no deliveries have been ordered with this specific due date.

$$q_{T-1j} = \sum_{k=0}^{j} q_{T-2k}\, b_{T-1,j-k} + \sum_{k=0}^{j} q_{T-1k}\, b_{N,j-k} \qquad 0 \le j < x \qquad (4.2.21)$$

During one 'cycle' there will be exactly one production period, so if we do not produce during the first $T-2$ periods, we will do it in one of the later periods.

$$q_{T-1x} = 1 - \sum_{i=1}^{T-2} q_{ix} \qquad (4.2.22)$$

By the formulae (4.2.17)-(4.2.22) the values of q_{ij} are determined. Now the expected costs between two production periods are given by:

$$C(x,T) = \sum_{i=1}^{T-1} \sum_{j=0}^{x} q_{ij} c_{ij} \qquad (4.2.23)$$

The average costs of the production rule for this pair (x,T) are now given by:

$$g(x,T) = \frac{C(x,T)}{T(x,T)} \qquad (4.2.24)$$

To determine the optimal pair (x,T), it is not necessary to determine the value of $g(x,T)$ for all possible pairs (x,T). We can limit the number of pairs we have to consider, making use of the following two properties:

Property 4.1 : for a given value of T the optimal value of x satisfies:

$$x \le \left\lceil \frac{g(x,T)}{p} \right\rceil + 1. \qquad (4.2.25)$$

In Property 4.2 we want to give limits for the optimal T-value for a given value of x. Let $q_{ij}(K)$ be the q_{ij}-value based on the (x,K)-rule and let $p_{ij}(K)$ be the probability that between two production periods the state (i,j) is visited, if we do not limit the state space in the *time*-direction. For these $p_{ij}(K)$ we thus allow $i \ge K$. Let $\bar{c}_K$ is the average cost in the K-th period after the last production period, if we use the $(x,K+1)$-rule and if we do not produce during that period.

Property 4.2 : for a given value of x and an arbitrary upper bound g^* for the average costs $g(x,T)$, the optimal value of T is less than or equal to K if for K the following holds:

$$\sum_{j=0}^{x-1}(g^*-\bar{c}_K)p_{Kj}(K) < h\,Ke_{K+1} \tag{4.2.26}$$

Now we will give the proofs of both properties.

Proof property 4.1 : We want to show that if y does not satisfy (4.2.25) for a given T, (y,T) cannot be optimal. The expected costs between two production periods in the $(y-1,T)$-rule, $C(y-1,T)$, are less than $C(y,T)$, because in the $(y-1,T)$-rule the penalty costs are less than in the (y,T)-rule and maybe the holding costs are also less, because in the $(y-1,T)$-rule we can produce earlier. Expressed in the q_{ij}'s associated with the pair (y,T) we have

$$C(y-1,T) = C(y,T) - p\sum_{i=1}^{T-1}(y-1)\,q_{iy-1} - h\sum_{i=1}^{T-2}q_{iy-1}\Big(\sum_{j=i+1}^{T-1}e_{i+1}\Big) \tag{4.2.27}$$

and

$$T(y-1,T) = T(y,T) - \sum_{i=1}^{T-1}q_{iy-1} \tag{4.2.28}$$

If $p\,(y-1) > g\,(y,T)$, that is y does not satisfy (4.2.25), we have that

$$g(y-1,T) = \frac{C(y-1,T)}{T(y-1,T)} \le \frac{C(y,T) - p\sum_{i=1}^{T-1}(y-1)\,q_{iy-1}}{T(y,T) - \sum_{i=1}^{T-1}q_{iy-1}}$$

$$< \frac{T(y,T)\,C(y,T) - C(y,T)\sum_{i=1}^{T-1}q_{iy-1}}{T(y,T)\,\big(T(y,T) - \sum_{i=1}^{T-1}q_{iy-1}\big)} =$$

$$= \frac{C(y,T)}{T(y,T)} = g\,(y,T)$$

We see that if y does not satisfy (4.2.25), $y-1$ yields lower average costs. Therefore y can only be optimal if it satisfies (4.2.25). □

Note: A result that follows immediately from Property 4.1. is the following. Let $g(y,T)$ be the average costs of the (x,T)–rule for an arbitrary pair (y,T). Since for the optimal value x the average costs $g(x,T)$ will be smaller than or equal to $g(y,T)$, we also have that the optimal value x satisfies:

$$x \leq \left\lceil \frac{g(x,T)}{p} \right\rceil + 1 \leq \left\lceil \frac{g(y,T)}{p} \right\rceil + 1. \tag{4.2.29}$$

This result can be used to determine the optimal x value.

Note: For $T \leq 2$ we also have that if $(y-1)p < g(y,T)$ then

$$g(y-1,T) = \frac{C(y,T) - p \sum_{i=1}^{T-1} (y-1)\, q_{iy-1}}{T(y,T) - \sum_{i=1}^{T-1} q_{iy-1}}$$

$$> \frac{C(y,T)\,(T(y,T) - \sum_{i=1}^{T-1} q_{iy-1})}{T(y,T)\,(T(y,T) - \sum_{i=1}^{T-1} q_{iy-1})} = g(y,T) \tag{4.2.30}$$

This implies that for $T \leq 2$ the optimal value of x satisfies:

$$x = \left\lceil \frac{g(x,T)}{p} \right\rceil + 1. \tag{4.2.31}$$

Proof property 4.2: We want to show that the pair $(x,K+1)$ cannot be optimal if 4.2.26 holds for a certain value of K. The difference between the expected time between two production periods for the $(x,K+1)$-rule and the (x,K)-rule is written as:

$$b := T(x,K+1) - T(x,K) = \sum_{i=K-1}^{K} \sum_{j=0}^{x} q_{ij}(K+1) - \sum_{j=0}^{x} q_{K-1j}(K) \tag{4.2.32}$$

$$< \sum_{j=0}^{x-1} p_{Kj}(K+1) \tag{4.2.33}$$

because $\sum_{j=0}^{x-1} p_{ij}(K) > \sum_{j=0}^{x-1} p_{i+1j}(K+1)$ for all $i \geq K$.

Let $\bar{c}_K$ be the expected costs in the K-th period after production if we use the $(x,K+1)$-rule and if we do not produce during that period. In later periods the expected costs per period, if we still do not produce, will be higher than $\bar{c}_K$. Therefore the following inequality holds for the total expected costs of the $(x,K+1)$-rule :

$$C(x,K+1) > C(x,K) + h\,Ke_{K+1} + b\bar{c}_K. \tag{4.2.34}$$

Then

$$g(x,K+1) = \frac{C(x,K+1)}{T(x,K+1)} > \frac{C(x,K) + h\,Ke_{K+1} + b\bar{c}_K}{T(x,K)+b} \tag{4.2.35}$$

Suppose g^* is an arbitrary upper bound for minimal average costs of the optimal (x,T)–rule, for instance the average costs of the (x,T)–rule for another pair. If 4.2.26 holds, we have that

$$\sum_{j=0}^{x-1} p_{Kj}(K+1) < \frac{h\,Ke_{K+1}}{g^* - \bar{c}_K} \tag{4.2.36}$$

As a result of (4.2.33)

$$b < \frac{h\,Ke_{K+1}}{g^* - \bar{c}_K} \tag{4.2.37}$$

and thus

$$b\bar{c}_K + h\,Ke_{K+1} > bg^* \tag{4.2.38}$$

Therefore

$$g(x,K+1) > \frac{C(x,K) + bg^*}{T(x,K)+b} \tag{4.2.39}$$

Now there are two possibilities:

1) $g(x,K) \geq g^*$, resulting in:

$$g(x,K+1) > \frac{g^*(T(x,K)+b)}{T(x,K)+b} = g^* \tag{4.2.40}$$

2) $g(x,K) < g^*$, resulting in:

$$g(x,K+1) > \frac{C(x,K)(T(x,K)+b)}{T(x,K)(T(x,K)+b)} = \frac{C(x,K)}{T(x,K)} = g(x,K) \tag{4.2.41}$$

and clearly the pair $(x,K+1)$ cannot be optimal. □

Now, to find the optimal pair (x,T), we can start for instance with $g^* = s$ and with $x = [s/p] + 1$. Starting with $K = 1$, we determine the first K-value for which (4.2.26) holds. Then we set $T = K$ and we determine $g(x,T)$ for this pair using (4.2.24) and we set $g^* = g(x,T)$. For this T value we determine the optimal x value. Therefore we use (4.2.29) and we choose $x = [g^*/p]+1$. If this is a new x value, then we determine $g(x,T)$, set $g^* = g(x,T)$ and we choose the next x-value. If the x value is not changed by the new choice, then we decrease the x value by one and determine $g(x,T)$. This is repeated until a further decrease yields a higher $g(x,T)$ value. Now we have determined the best x value for the first choice of T.

Let g^* be the minimal value of $g(x,T)$, that is determined until now. For the x value associated with this minimal value we will now determine the optimal T value. We can start with $K=T$ and decrease K until (4.2.26) no longer holds. Then we give T the minimal value of K for which (4.2.26) holds. We determine $g(x,T)$ and decrease T until a further decrease yields a higher $g(x,T)$ value. Now we have determined the best T value for the second choice of x. If the new T value is different from the first T value, we determine a new best x value in the same way as the best x value for the first choice of T. If this yields another x value, we continue with determining the new best T value. This procedure is repeated until a further decrease of x or T yields higher costs. Determining the average costs for a new pair (x,T) is usually quite simple, since most of the q_{ij}-values are not affected by decreasing one of the two components, x or T, by one unit.

An alternative starting pair (x,T) can be found by considering the cyclic production rule π. Using (4.2.10) we determine T and g_π. Then the starting value for x will be $[g_\pi/p]+1$ and the starting value for T will be the same value as the T-value found in (4.2.10). The ultimate optimal value of x will be smaller than or equal to this starting value, assuming that for the optimal pair we have average costs $g(x,T)\le g_\pi$, but the optimal value of T can also be larger than the starting value. This can be checked by using (4.2.26).

4.2.5. Numerical results

The average costs of the optimal policy can be determined by means of dynamic programming. The average costs of the Wagner-Whitin approach and the average costs of the Silver-Meal approach can be determined by calculating the probabilities for all possible orders state vector. Just as the computation of the average costs of the optimal policy, this may take a lot of computational efforts, because the state space will usually be very large (see Dellaert (1986)). Sometimes we can obtain a considerable reduction of the computational efforts if some of the components of the order state vector can be ignored. This may occur in the Wagner-Whitin approach, but it is more likely to happen in the Silver-Meal approach, if we always produce only orders with a residual lead time that is smaller than N. Despite this reduction, for problems of a normal size the computation will still be rather difficult. The average costs of the optimal (x,T)-rule can be determined with very little effort, especially if we use the properties 4.1 and 4.2.

In order to study the performance of each of the rules, we will determine the average costs for some examples. The size of the examples is small, so we can compute the costs of each production rule. We will consider the following rules:

OPT, the optimal policy;

SM, the Silver-Meal-like production rule;

WW(H), the Wagner-Whitin-like production rule with H periods;

XT, the optimal (x,T)-rule;

CYC, a cyclic production rule, in which we produce every T periods if there is any demand for the first T periods.

We have considered two sets of simple examples. In the first set we have a maximum lead time $N=4$, penalty costs $p=3$, holding costs $h=1$ and a binary demand $d_{i1}=1-d_{i0}=d$ for i=1,..,4. For some different values of d and s we have the following results:

d-value	0.25	0.25	0.50	0.50	0.75	0.75
s-value	3.25	8.00	6.50	16.00	9.75	24.00
rule	average costs per period					
OPT	1.8895	3.7147	4.5357	8.1705	7.0425	12.6002
SM	1.9002	3.7173	4.5392	8.1793	7.0445	12.6054
WW(1)	1.8954	3.7830	4.5384	8.2135	7.0432	12.6054
WW(2)	1.8940	3.7770	4.5384	8.1910	7.0432	12.6020
WW(3)	1.9002	3.7815	4.5438	8.1861	7.0425	12.6020
WW(4)	1.9002	3.7815	4.5438	8.2002	7.0425	12.6020
XT	1.9074	3.7326	4.5392	8.1965	7.0451	12.6125
CYC	2.2123	4.1655	4.7373	8.4987	7.1249	12.7500

Table 4.1 The average costs of the rules in the first set of examples.

In the second set of examples we have a maximum lead time $N=2$, penalty costs $p=3$, holding costs $h=1$ and a Poisson-distributed demand with parameter 1 for both lead times. For three different values of the set-up costs s the results are given in Table 4.2.

s-value	7	8	9
rule	average costs per period		
OPT	4.8498	5.3335	5.8113
SM	4.8537	5.3365	5.8261
WW(1)	4.8574	5.3335	5.8258
WW(2)	4.8574	5.3335	5.8258
XT	4.8767	5.3807	5.8847
CYC	5.4359	5.9267	6.4176

Table 4.2 The average costs of the rules in the second set of examples.

From these examples one might get the impression that the SM-rule is always better than the optimal (x,T)-rule. This however, is not the case: if in the second set of examples we take a Poisson-distributed demand with parameter 1.01 and set-up costs $s=9$ then the average costs of the SM-rule are 5.9308, whereas the average costs of the optimal (x,T)-rule are 5.9082.

4.2.5.1. Conclusions

The examples showed us that the heuristic rules perform very well, especially compared with the fixed production cycle rule. The difference in average costs between the heuristics and the optimal policy is always less than a few percent. It might be interesting to see how the differences between the optimal policy and the heuristics arise. Observations of the examples that have been described, learned us the following elements:

1) in the SM-rule we produce less often than in the optimal policy: we delay production too long and if we produce, we produce too much;

2) the (x,T)–rule contains the same elements as the SM-rule, but usually even stronger;

3) in the WW-rule we delay production too long and if we produce, we produce less than the optimal amount.

4) in the cyclic rule we produce too often; even in the situation in which r_1 equals 0.

Now which of the heuristic approaches must be preferred in more practical or in more complex situations? The Wagner-Whitin algorithm as it is used for the deterministic situation suffers already from a lack of acceptance, merely because of its complexity. The WW-rule we have considered is even more complex, so this will never be the ideal approach. The decision rules for SM are not complex at all and its performance is usually slightly better than the best (x,T)–rule. The disadvantage of the SM-rule is that we are usually unable to compute the average costs of this production rule. This makes it less useful for situations in which we have to take decisions about long-term agreements with clients or other problems in which we are interested in the average costs. In those situations one would need simulation to get a reasonable insight into the average costs.

The (x,T)–rule however, offers a very simple decision rule with known average costs. Especially if we want to analyse or change a situation, the (x,T)–rule must be preferred. As a basis for a production rule in a more complex situation the SM-rule and (x,T)–rule can both be useful, as we will see later on. The use of the WW-rule in more complex situations will be restricted to an extended version of the WW(1)-rule. In Dellaert (1987) some examples of the use of the (x,T)-rule for making agreements with clients are given. In that paper we also present simple rules for the uncapacitated situation to improve the (x,T)–rule until it is nearly optimal and a method to estimate the average costs of the SM-rule.

4.2.6. Orders with different priorities

4.2.6.1. Introduction

Until now, we have assumed that all orders were of equal importance. In a practical situation, this is not always the case. Some clients may find a shorter lead time more important than others and therefore willing to pay a higher price. Other clients may be content with a longer lead time, but nevertheless they will not allow orders to be delivered too late. Obviously, orders for which different prices have to be paid, will be treated differently. In Chapter 6 orders with different priorities will have different average delivery times and in Chapter 7 the proposed lead times will be different. In this subsection we will consider a simple way for the different treatments. Orders with different priorities will have different penalty costs, but equal holding costs. If the holding costs are proportional to the penalty costs, for the different priorities, we can simply replace an order with a high priority by a number of orders of the lowest priority. In this proportional costs situation, no changes would be necessary in the production rules described so far. In this subsection we will assume that the holding costs are not proportional to the penalty costs. We will observe that the Silver-Meal approach and the (x,T)–rule can easily be adapted to this kind of situation.

4.2.6.2. Model

In Chapter 2, we have listed a number of different elements of the penalty costs. Part of these costs can be considered to be real costs, but another, often much larger part, consists of speculative elements or simply represent management policies. Therefore it will be no problem to choose the penalty costs for orders with different priorities as an integer number multiplied by p. For example, in a certain situation for some of the orders the penalty costs will be p, for some other orders $2p$ and for another category of orders $3p$ or $4p$. In the production rules which we have considered until now, we base the decisions upon the state vector $r=\{r_1,..,r_N\}$. If we distinguish orders with for instance three different priorities, a possible state description is given by denoting the number of orders of priority j with a residual lead time of i periods by r_{ij}. This would imply that the state vector contains $3N$ components and this will soon lead to an immense state space.

Due to the increased state space, the use of the Wagner-Whitin approach and the use of the optimal policy will become much more complex, because in these production rules we have to consider the order state during a number of periods. In the Silver-Meal approach and in the (x,T)–rule we only consider the action for the first period. In order to determine the penalty costs, if we do not produce during this first period, we have to know the sum of the r_1 values for the different priorities weighted with the penalty costs per order. In case of a set-up, the holding costs do not depend on the number of orders for each type separately. Therefore, for the decisions in both the Silver-Meal approach and in the (x,T)–rule, we can consider the aggregated order state vector, where the first component is the weighted sum of the orders for the different priorities and the other components simply are the sum of the orders for the different priorities with a certain residual lead time.

The weighting of the orders that contribute to the first component of the aggregated order state vector can be done in the following way. We define the *penalty points* of an order as the penalty costs for that order divided by p, so the penalty points may be 1,2,3 etc. Now we define r_1' as the sum of the penalty points of those orders that have to be produced by the end of the period or in previous periods. The other elements of the order state vector, $r_2',..,r_N'$, are defined as the sum of the number of orders with a particular residual lead time. For instance, r_2' is the sum over all priorities of the number of orders with a residual lead time of 2 periods. The description of the transition probabilities for this order state vector r' will be slightly more complex, but on the other hand, there is a large state space reduction.

The new state vector r' does not provide the necessary information for a good use of the optimal policy and for the Wagner-Whitin approach. For these rules it is indeed necessary to know the order state vector for each of the different priorities separately. The (x,T)–rule and the Silver-Meal approach can be used straightforwardly in this situation. Of course the fixed cycle production rule π (cf. (4.2.10)) can also be used without any complications. The use of these rules is now exactly the same as described in 4.2.4.2. and 4.2.4.4.

4.3. A fluctuating demand rate

4.3.1. Introduction

In the previous section, we assumed that orders that are delivered too late had no influence upon future demand. The effect of unsatisfied clients was expressed in the penalty costs p, but not in the demand. This way of modelling unsatisfied clients is very common, and it can be found in many production-inventory models. Nevertheless, in a practical situation orders that are delivered too late may have an influence upon the future demand. In this subsection, we consider a situation in which there is a finite population of clients. The number of clients in the population is subject to changes due to the lateness of some of the orders and also to other changes which have nothing to do with the lateness. We will assume that we have fixed lead times, so we do not need a rule for the lead times. The main element of interest to us is to see whether or not it should be allowed to replace a situation with a fluctuating number of clients by a stationary demand distribution and to estimate the loss of orders due to unsatisfied clients by means of penalty costs. There are two important reasons for our interest in this replacement: we are interested in the behaviour of the (x,T)–rule and the SM- rule in a situation in which the demand distribution is not completely known and we are interested in the error in the results of the analysis of the (x,T)–rule in such a situation.

4.3.2. Model

The situation in which the population of clients is subject to different kinds of changes, as described above, is modelled in the following way:

- clients can leave the population, independent of the lateness of the orders, according to a stochastic process where the number of clients, in other words the size of the population, is one of the parameters.
- clients can leave the population, depending on the lateness of the orders, according to a stochastic process where the number of late orders is one of the parameters.
- new clients can join to the population according to a stochastic process that is independent of the size of the population and of the number of late orders.

Let $N(t)$ be the population size at the beginning of period t, $D(t)$ the decrease in the population size during period t, which is independent of the lateness of the orders, $L(t)$ the lateness-dependent decrease in the population size and $I(t)$ the number of new clients during period t. The stochastic processes are independent of each other, except that in every period the total decrease cannot exceed the number of clients in the population. Therefore the population size in period $t+1$ cannot be less than the number of new clients during period t. Hence the number of clients in the population can be described by the following formula

$$N(t+1)=\max(N(t)-D(t)-L(t)+I(t), I(t)), \qquad (4.3.1)$$

For a firm, the population size is not that important. Much more important is the resulting demand by these clients. In our model, all clients have the same properties with respect to leaving, arriving as a new client and with respect to the demand they order. This implies that the distribution of the demand for a certain period, having n clients is the n-fold convolution of the distribution of the demand with 1 client. We will assume that every client can order products in every period for each of the N possible lead times.

For a given distribution of the demand with 1 client and given distributions for the stochastic processes $D(t)$, $L(t)$ and $I(t)$, we have to determine the action as a function of the population size $N(t)$. The object should be the maximisation of the profit. The profit is given by the revenues of the orders minus the costs for set-ups and the holding costs. We do not have penalty costs in this situation, since the costs of lateness are now expressed in the demand and the number of clients.

In order to determine the best actions in every situation we will consider the two heuristics we preferred in Section 4.2.: the Silver-Meal-like production rule and the (x,T)–rule. The use of the optimal policy becomes much more complicated than in the previous section, because of the introduction of a more complicated state which has to include the population size and therefore more complicated transition probabilities, all together leading to a production rule which can hardly be of practical use. In the heuristics, the penalty costs again are a basic element. The penalty costs are now defined as the expected volume of missed revenues as a result of departed customers if an order is delivered one period too late. If the revenues and the stochastic processes which govern the decrease and increase of the population size are known, the penalty costs can be estimated in a simple way and the heuristics can be used in the same way as in Section 4.2. We will explain this by studying an example.

4.3.3. Example of lateness-dependent demand

Suppose that the probability that an arbitrary client leaves at the end of a period, independent of the lateness of any of his orders, equals α. Suppose that any late order leads to a departure of one of the remaining clients with probability β. The reason for this second assumption is that it simplifies the analysis without having a large effect upon the results. We assume that all demand ordered will also be delivered, independent of a possible departure of the client. New clients will arrive according to a Poisson process, with intensity λ per period, independent of any of the other stochastic processes. Writing $E(t)$ as the number of late orders during period t, we have the following formulae for $D(t)$, $L(t)$ and $I(t)$:

$$\boldsymbol{P}\{D(t)=n\}=\binom{N(t)}{n}\alpha^n(1-\alpha)^{N(t)-n}, \qquad n=0,1,..,N(t) \qquad (4.3.2)$$

$$\boldsymbol{P}\{L(t)=n\}=\binom{E(t)}{n}\beta^n(1-\beta)^{E(t)-n}, \qquad n=0,1,..,E(t) \qquad (4.3.3)$$

$$\boldsymbol{P}\{I(t)=n\}=\frac{\lambda^n}{n!}e^{-\lambda} \qquad n\in \boldsymbol{N} \qquad (4.3.4)$$

From (4.3.2) we can learn that an arbitrary client will stay on average α^{-1} periods if lateness does not occur. The process of leaving is memoryless, for geometrical, implying that if a client decides to stay for one more period, the expected time of his stay is again α^{-1} periods. If however, an order is delivered one period too late, then the expected staying time for one of the clients will decrease to $(1-\beta)/\alpha$, leading to a loss of revenues during β/α periods. The penalty costs p, used in the heuristics, will be determined according to this loss.

Suppose that in the situation in which there is exactly one client, the probability that during an arbitrary period t the demand for period $t+i$ equals j is given by d_{ij}, $i=1,2,..,N$. The expected demand for each of the periods $t+i$ is given by

$$u_i=\sum_{k=1}^{\infty}k\,d_{ik} \qquad i=1,2,..,N \qquad (4.3.5)$$

and the average amount ordered per period per client is given by

$$\hat{u}=\sum_{i=1}^{N}u_i \qquad (4.3.6)$$

If the revenues per ordered item are given by w, then the expected revenues per period per client are $w\hat{u}$. The penalty costs p, being the expected loss of revenues for one late order during one period, are then given by $w\beta\hat{u}/\alpha$. This enables us to describe the penalty function $p(a)$, which can be used in the Silver-Meal approach. This function has been defined in (4.2.7) and (4.2.8). For a situation with n clients we have:

$$p(a)=\frac{w\beta\hat{u}}{\alpha}\sum_{i=2}^{a}(a+1-i)\sum_{j=1}^{i-1}nu_j \qquad \text{for } a=2,3,..,N \qquad (4.3.7)$$

$$p(a)=0 \qquad \text{for } a=0,1 \qquad (4.3.8)$$

4.3.4. Numerical results

Now that we have found alternative values for the penalty costs and the penalty function, we can use the heuristics from the previous section again. In this subsection we will demonstrate the use of the Silver-Meal approach and of the (x,T)–rule and compare the results of these rules with a fixed cycle rule. Therefore we will do a simulation for two examples, in which we assume that the population size is known at the end of every period. For these two examples we will also consider the (x,T)–rule and the SM-rule in combination with the assumption that the population size is not known at the end of a period. In this situation with incomplete information, we assume that the long-term average population size is known. Furthermore we will compare the results with two different ways of analysis: the analysis of the (x,T)–rule with the incomplete information about the population size and the analysis of the (x,T)–rule with the exact knowledge of the population size. The computation time of the analysis of the (x,T)–rule with the exact knowledge, depends on the maximum number of clients in the population and on the penalty costs. If the penalty costs are large enough the decision variable x in the (x,T)–rule will be rather small. First we will describe the use of both heuristics in the situation with a fluctuating population size.

The Silver-Meal approach can be used in the same way as described in 4.2.4.2., that is: if we observe a order state vector $r \in R$, we take that action a for which

$$\frac{q_r^a + p(a)}{a + \delta(a)} \tag{4.3.9}$$

is minimal over $a \in A(r)$. Notice that in (4.3.9) the penalty function $p(a)$ is the only element that depends on the population size. In the simulation of the Silver-Meal approach with the incomplete information, we will replace the element $N(t)$ in the penalty function by a constant $\overline{N}$.

The (x,T)–rule in itself is not much different from the one described in 4.2.4.4. Now for every number of clients n, the probabilities for the demand, $b_{ij}(n)$ are different and therefore the optimal set of $(x(n),T(n))$ has to be determined for every possible value of n and then stored. The use of the (x,T)–rule becomes very simple: for a given value of n we have to check whether the number of orders that would be too late, if we do not produce during the period, is at least equal to $x(n)$ and if so, we will produce all the required deliveries for the first $T(n)$ periods.

Using the method described in (4.2.17)-(4.2.22) for certain values of n and $(x(n),T(n))$, we can determine the expected number of visits q_{ij} during one production cycle to each of the states (i,j). By dividing these values q_{ij} by $T(x(n),T(n))$, the average length of one production cycle, we obtain the probabilities p_{ij} to be in state (i,j), $1 \le i \le T-1, 0 \le j \le x$. From these probabilities we can determine the probability that the number of late orders during an arbitrary period t, in which we have n clients, equals j:

$$\boldsymbol{P}\{E(t)=k\} = \sum_{i=1}^{T-1} p_{ik} \qquad k=1,..,x-1 \qquad (4.3.10)$$

These probabilities for $E(t)$ and the probabilities for a normal departure and for joining the population can be used to determine the distribution of $N(t)$, the population size in an arbitrary period. In the (x,T) analysis we will determine this distribution and also the average costs and revenues per period. This analysis can only be performed under the assumption that the number of late orders in subsequent periods is independent. We will also perform the analysis of an (x,T)–rule in which the population size is fixed. In Table 4.3. this analysis is denoted by the term 'inc. (x,T)-analysis', where inc. stands for incomplete information, or in other words, we do not use the exact information about the population size to determine the action.

In Table 4.3. we will also give the simulation results for the (x,T)–rule and the Silver-Meal approach, once with the decisions based on the real population size and once with the decisions based on the average population size. The simulation of the production rules with the decisions based on the average population size will be denoted by 'inc. (x,T) simulation' and 'inc. SM simulation' respectively. We will also perform a simulation with the cyclic strategy π, in which we produce every T periods.

In Table 4.3. we will give the results for two examples. In both examples we have used a normal departure probability α=0.01, a departure probability for a late order β=0.20, a maximum lead time N of 4 periods, set-up costs 50, holding costs 1 per unit per period and revenues of 10 per item ordered. In both examples the demand per client per lead time is Poisson distributed. In Example 1, the demand rates are $u_i = 0.01$ for the lead times $i = 1,..,4$ and in Example 2 the demand rates are $u_i = 0.01$ for the lead times $i = 1,2,3$ and $u_4 = 0.02$ for a lead time of 4 periods. These demand rates result in penalty costs $p = 8$ in Example 1 and $p = 10$ in Example 2. The values for the average population size have been chosen according to results of the simulation of the two production rules with the exact knowledge of the population size, which has yielded an average $\overline{N}$ of 82 in Example 1 and 87.5 in Example 2. In both examples, the fixed cycle production rule implied that we produce every three periods.

rule	N(t)-average		costs		profit	
(*x*,*T*)-analysis	82.60	87.53	19.00	23.07	14.04	20.70
(*x*,*T*)-simulation	82.19	87.82	18.93	23.31	13.94	20.61
SM-simulation	82.10	87.53	18.86	23.11	13.98	20.66
inc. (*x*,*T*)-analysis	82.00	87.50	18.86	23.59	13.94	20.16
inc. (*x*,*T*)-simulation	82.38	87.58	19.03	23.18	13.93	20.62
inc. *SM*-simulation	82.44	88.53	19.02	23.73	13.96	20.54
fixed cycle	78.72	79.56	18.50	19.31	12.99	20.47

Table 4.3 Results with left Example 1 and right Example 2

From Table 4.3. we can learn that the difference between the analysis and the simulation of the rules with the complete information, which is partly due to the assumption of independent $E(t)$'s and partly due to the inaccuracy of the simulation, is very small. In the analysis of the (*x*,*T*)–rule with incomplete information the average costs per period and the profit are quite near to the values found for the rules with the complete information. This implies that some lack of information, or some wrong information about the demand will not be very harmful for the results of the analysis of the (*x*,*T*)–rule. The same holds for the use of the (*x*,*T*)–rule and also for the use of the Silver-Meal approach. For both examples the difference between the rules with a full knowledge of the population size and the rules with incomplete information are very small and for example, much smaller than the difference between the rules and the fixed cycle production rule. We can also see that the profit in the fixed cycle production rule is less than the profit in the other rules.

4.4. A decision rule for lead times

4.4.1. Introduction

Clients can react upon the accuracy of the due dates, but also upon the length of the promised lead times. If an order is delivered too late, the client can withdraw the order or stop ordering for some time. Usually the order will not be withdrawn and thus the reaction of the clients towards the accuracy of the due dates can be considered as a long-term process. The reaction of a client upon the length of the promised lead time can be much more direct. If a client is not content with the lead time he can address another company, where he may obtain the required lead time. In this section we will consider firm-initiated lead times. The lead times for orders are determined at the end of a period, followed by a decision about the production, where all the orders that arrived during a period obtain the same lead time. We assume that all orders have the same priority. The firm-initiated lead time depends on the number of previous orders which have not been produced, and on the residual lead times of these orders and also on the number of new orders placed during the period. Some of the clients may not be content with the proposed lead time and therefore they may withdraw the order. In this situation, we want to maximise the profit by generating the best lead times.

4.4.2. The model

The firm offers lead times from 1 period up to N periods. The probability that a client accepts a certain lead time k, is denoted by A_k, $1 \leq k \leq N$, and these probabilities are known to the firm; they are the same for all clients. The withdrawal of an order will have no effect upon future orders of a client and we will assume a stationary process of ordering. The probability that the number of orders in an arbitrary period equals i is denoted by d_i, with $0 \leq i \leq M$. E denotes the average number of orders. We want to concentrate on the decision rule for the lead times and therefore we want to simplify the production planning as much as possible. It will be required that all accepted orders are produced in time, thus avoiding penalty costs and unsatisfied clients. Sometimes orders will be produced too early and this will lead to holding costs h per order per period. Again, the set-up cost is given by s, the revenues per order by w and there are no capacity restrictions.

First, we shall consider a cyclic production rule in combination with a cyclic rule for the lead times and determine the average profit of this approach. Then we will give a description of a dynamic programming rule (DPR) for the lead times. This DPR is based te reduction and the decision rule itself is determined by means of dynamic programming. We will conclude this section with the description of a simplified version of the DPR, which will be denoted as the SDPR. Then some examples will be given to compare the performance of the different decision rules.

4.4.3. A cyclic approach

A simple indication of the profit can be obtained by applying a cyclic production rule to the situation and by basing the lead times for the orders upon the production plan. We will consider the production rule π where we produce every T periods if there is any demand. Orders that arrive k periods before the next scheduled production period will have a proposed lead time of k periods, $1 \le k \le T$. Because of the knowledge of the production schedule, the due dates will be exact and there will be no holding costs. Every T periods we will have set-up costs, at least if there is any demand. The probability that there are no orders to be produced after the cyclic interval T will be denoted by p_{0T}, which can be written as

$$p_{0T} = \sum_{k=1}^{T} \sum_{j=0}^{M} d_j (1-A_k)^j \qquad (4.4.1)$$

The average profit of the approach is now given by

$$g_\pi = \frac{\sum_{i=1}^{T} wEA_i - s(1-p_{0T})}{T} \qquad (4.4.2)$$

where the length of the cyclic interval T is chosen in such a way that it yields the maximum for g_π. Of course, this approach will not be the optimal approach, but it offers an indication of the possible profit and it can also be used for a comparison with other rules.

4.4.4. The dynamic programming rule

At the end of every period, we have to take a decision about the lead times for the new orders that have been arrived during the period. Then the clients can accept these lead times or withdraw the order. If the process of acceptance has taken place, still at the end of the period, a decision about the production has to be made. The production rule for this situation can be rather simple. If there are no required deliveries for the first period, then it makes no sense to produce already and therefore action $a = 0$ is the only reasonable option. Because of the assumption that all accepted orders will be produced in time, we have to produce if there are required deliveries for the first period. Usually most of the orders will have about the same due date. A reasonable action in this situation is to produce all orders, in other words action $a = N$, where N is the maximal lead time of an order. However, we can also think of a situation in which there are clearly two groups of orders: one group with a residual lead time of one period and another group of orders, which is not too small, with a residual lead time of three or more periods. In this situation the most reasonable action can be the production of only the orders that have to be delivered by the end of the first period, or in other words: take action $a = 1$. Of course, we can only come to a situation like this if the rules for the lead times make an occurrence of such a situation possible.

If we use this production rule, the two most important elements for the rules for generating lead times are the total number of orders and the minimal value of the residual lead times of the accepted orders. If the lead time for the new orders is the same as the minimal value of the residual lead times, the lead time decision does not imply extra holding costs. If the lead time for the new orders is shorter than the minimal value, then we have to pay holding cost for the difference in periods. In the situation in which we decide only to produce the new orders, as far as they are accepted, in the first period, then we do not have holding costs. From this discussion it follows that the original state space, with the state space vector $r=(r_1,r_2,\ldots,r_N)$, where r_i denotes the number of other with a residual lead time of i periods, can be limited to the two most important elements: the total number of accepted orders and the minimal values of the residual lead times of the accepted orders. We will denote a state by (y,t), where y is the total number of orders, $y \in N$, and t the number of periods, $0 \leq t \leq N$. The state in which there are no orders will be denoted by $(0,N)$. In the states with $t = 0$ we have produced all orders during the period.

Now we come to the description of the DPR. In this rule we make the decision about the lead times and the action for the production simultaneously. In a situation where we have decided to produce and where all orders for the first period have been withdrawn, the decision to produce will be cancelled. If no new orders have arrived during a period, the possible actions for the production will be $a=N$ if there are required deliveries for the first period, or $a=0$ otherwise. In the DPR we have to make a choice for every state (y,t) in combination with j new orders. This choice will be denoted by $Ch(y,t,j)$. If there are no new orders during a period, no decision has to be made. The value of $Ch(y,t,j)$ will be the offered lead time to the j new orders, if this lead time is 2 periods or more. If the offered lead time to the j new orders is equal to 1, then we will write $Ch(y,t,j)=1$ if the lead time decision is combined with the decision to produce all orders, or $Ch(y,t,j)=0$ if the lead time decision is combined with the decision to produce only the new orders, as far as they are accepted. In the states (y,t) with t=0,1,2, the production rule does not allow the choice $Ch(y,t,j)=0$. If the choice $Ch(y,t,h)=0$ is made, the cost effect of the new orders will be calculated immediately. Therefore, the effect of this choice upon the future periods is the same as the effect of having no new orders during a period.

In order to make the correct choices $Ch(y,t,j)$, we will use a value function $v(y,t)$ which denotes the marginal profit until the next production period. In the DPR we choose this value function similar to the relative cost function $v(r)$ in (4.2.6):

$$v(r)=\min_{actions}\left[q_r^a+\sum_z P_{rz}^a\, v(z)-g\right]$$

There are some differences between our $v(y,t)$ and this relative cost function $v(r)$. Instead of considering the possible 'actions', we will consider the possible choices. We will maximise the profit instead of minimise the costs. Due to our state space, the holding costs have to be paid as soon as it becomes clear that an order will be produced earlier. If the holding costs have been paid, we can treat all orders as if they have the same due date. The revenues will be received in the periods in which the orders are produced, that is in the states $(y,0)$. In the same states $(y,0)$ we will pay the set-up costs. An exception is the choice $Ch(y,t,j)=0$. If this choice is made, the revenues for the orders that will be produced in the first period and the set-up costs will charged immediately. Let g_{dpr} denotes the average profit per period if we use the optimal DPR. The value of g_{dpr} is not known at the beginning and will be determined in an iterative way by using a policy iteration method, starting with an arbitrary value g_0.

Due to the cost structure where the holding costs are paid immediately, where the revenues are received in the production period and where the set-up costs are paid in the production period, we have the following direct costs. If in the state (y,t) j new orders have arrived and the choice Ch is made, the 'direct costs' of this choice, analogue to the direct costs q_r^a in (4.2.6), will be denoted by $dc(y,t,j,Ch)$. The value of $dc(y,t,j,Ch)$ is given by:

$$dc(y,0,j,Ch) = yw - s \qquad y \geq 1,\ j \geq 0,\ Ch \geq 0$$

$$dc(y,t,j,Ch) = h\,y \cdot (t-1-Ch) \qquad y \geq 1,\ j \geq 1,\ 1 \leq Ch \leq t-2$$

$$dc(y,t,j,0) = jwA_1 - s \qquad y \geq 1,\ 3 \leq t \leq N,\ j \geq 1$$

$$dc(y,t,j,Ch) = 0 \qquad \textit{otherwise}$$

The probability that the lead times for i out of j orders will be accepted if the proposed lead time equals l is denoted by P_{ij}^l. This is of course a function of the probability A_l that a client accepts lead time l for an order:

$$P_{ij}^l = \binom{j}{i} A_l^i (1-A_l)^{j-i} \qquad i=0,..,j;\ l=2,..,N \qquad (4.4.3)$$

Using this probability we can calculate the probabilities for moving to state (x,k) if in the state (y,t) j new orders have arrived and the choice Ch has been made. These probabilities will be denoted by $Q(y,t;x,k)$:

$$Q(y,t;x,Ch-1) = P_{x-y,j}^{Ch} \qquad x>y;\ j \geq 1;\ Ch=1,..,t$$

$$Q(y,t;y,t-1) = P_{0,j}^{Ch} \qquad y \geq 1;\ j \geq 0;\ Ch=1,..,t$$

$$Q(y,t;y,t-1) = 1 \qquad y \geq 1;\ j \geq 1;\ Ch=0;\ t \geq 2$$

$$Q(y,0;x,Ch-1) = P_{x,j}^{Ch} \qquad x \geq 1;\ j \geq 1;\ Ch=1,..,N$$

$$Q(y,0;0,N) = P_{0,j}^{Ch} \qquad j \geq 0;\ Ch=1,..,N$$

Now we can give the definition of $v(y,t)$:

$$v(y,t) = \sum_{j=0}^{M} d_j \left\{ \max_{Ch} \left[dc(y,t,j,Ch) + \sum_{x,k} Q(y,t;x,k) \cdot v(x,k) - g_0 \right] \right\} \qquad (4.4.4)$$

Since these values indicate relative profits, we can choose the value for one of the states freely, for instance:

$$v(1,0) = w - s$$

Due to this choice, we also have:

$$v(y,0) = yw - s \qquad y \in \mathbb{N} \qquad (4.4.5)$$

If a choice for g_0 has been made, for instance $g_0 = g_\pi$, (cf. (4.4.2)), we can determine all values $v(y,t)$, starting with t=1 and then increase t. In this process of calculating the $v(y,t)$ values, we also have found values for $Ch(y,t,j)$, where $Ch(y,0,j) = Ch(0,N,j)$ for all t and j. From the calculated $v(y,t)$ values we can determine a new value for g_0 and determine the optimal DPR by means of this value iteration. However, in this situation we will choose for the policy iteration method, because this method offers more information about the different cost elements. First we will consider the policy based on the choices $Ch(y,t,j)$. By determining the steady-state probabilities to be in state (y,t), given the rule based on these choices, we can learn the average time between two production periods, the average profits, set-up costs and holding costs per period. In an analogue way to 4.2.4.4. we define $q(y,t)$ as the average time spent in state (y,t) between two normal production periods. Normal production periods are defined as periods in which we produce all orders. The production periods in which we produce only the orders with a residual lead time of 1 period will be called *inbetweenies*, since this production is performed between two normal production periods. For notational reasons we introduce the set $V(y,t,k)$, the set of values for the demand, where in state (y,t) the choice $Ch(y,t,j) = k$ has been made:

$$V(y,t,k)=\left\{ j \mid Ch(y,t,j) = k \right\} \tag{4.4.6}$$

Now the probability to enter state (x,k) coming from state (y,t) is given by

$$\boldsymbol{P}\left\{(y,t) \rightarrow (x,k)\right\} = \sum_{j \in V(y,t,k+1)} d_j P^{k+1}_{x-y,j} \qquad x>y,\ k<t \tag{4.4.7}$$

Furthermore we have

$$\boldsymbol{P}\left\{(y,t) \rightarrow (y,t-1)\right\} = d_0 + \sum_{k=1}^{t} \sum_{j \in V(y,t,k)} d_j P^k_{0j} +$$

$$+ \sum_{j \in V(y,t,0)} d_j (1-P^1_{0j}) \qquad y>0,\ t>0 \tag{4.4.8}$$

$$\boldsymbol{P}\left\{(0,N) \rightarrow (0,N)\right\} = d_0 + \sum_{k=1}^{N-1} \sum_{j \in V(0,N,k)} d_j P^k_{0j} \tag{4.4.9}$$

and

$$\boldsymbol{P}\left\{(y,0) \rightarrow (0,N)\right\} = d_0 + \sum_{k=1}^{N-1} \sum_{j \in V(y,0,k)} d_j P^k_{0j} \tag{4.4.10}$$

With the use of (4.4.7)-(4.4.10) we can calculate the average time spent in each of the states (y,t) in a straightforward way and from these $q(y,t)$ values we can determine the elements of interest. The average time between two normal production periods, denoted by T, is simply the sum of all $q(y,t)$ values:

$$T = \sum_y \sum_t q(y,t) \tag{4.4.11}$$

The average number of inbetweenies between two normal production periods, denoted by Ib, can be found by considering the j values for which in the states (y,t) with $2 \le t \le N-1$ the choice $Ch(y,t,j)=0$ has been made and where at least one of the orders has not been withdrawn:

$$Ib = \sum_{y>0} \sum_{t=2}^{N-1} q(y,t) \sum_{j \in V(y,t,y+1,0)} d_j (1-P_{0j}^1) \tag{4.4.12}$$

The average storage costs between two normal production periods, denoted by H, can be found by considering the j values for which in the states $y,t)$, with $2 \le t \le N-1$ a choice $1 \le Ch(y,t,j) < t$ has been made and where at least one the orders has not been witdrawn. In this situation the holding costs are $h \cdot y \cdot (t - Ch(y,t,j))$. Hence H is given by:

$$H = h \sum_{y>0} \sum_{t=2}^{N-1} y\, q(y,t) \sum_{k=1}^{t-2} \sum_{j \in V(y,t,y+1,k)} d_j P_{0j}^k (t-1-k) \tag{4.4.13}$$

The expected revenues during one cycle is found by summing the expected number of accepted orders for each of the states:

$$R = w \left[\sum_{y>0} \sum_{t=0}^{N-1} q(y,t) \sum_{j=1}^{M} j d_j A_{Ch(y,t,j)} + q(0,N) \sum_{j=1}^{M} j d_j A_{Ch(0,N,j)} \right] \tag{4.4.14}$$

where we use for notational convenience $A_0 = A_1$. Now the profit per period, g_0, is given by:

$$g_0 = \frac{R - H - (1+Ib)s}{T} \tag{4.4.15}$$

With this value we can repeat (4.4.4) and determine g_0 again, until the rule does not change any more and we have found g_{dpr}. Usually this occurs after a few steps.

Note: the value of N, the maximum lead time that is offered, may be subject to changes, because of the changes in the $v(y,t)$ values. The value of N can be found by:

$$N = \min \{ n \in \mathbf{N} \mid v(0,n+1) = v(0,n+2) \} \tag{4.4.16}$$

4.4.5. Simplified dynamic programming rule

The calculation of g_{dpr} requires a lot of work, especially if N and M, the maximum number of orders per period, are quite large. Some of this work can be avoided by making some assumptions about the possible choices in a state, as we will do in the properties 4.3 and 4.4. Much more work can be avoided if we do not use the relative costs $v(y,t)$ as defined in (4.4.4), but a simplified version for these costs. In this simplified version, the values of $v(y,t)$ with respect to the expected profit until the first (normal) production no longer depend on the DPR, but on the cyclic rule π. Therefore, this profit does not change if the production rule changes during the iteration. The only element subject to changes in $v(y,t)$ is the average profit of this simplified rule, an element that can change as long as the rule changes. Thus we can write

$$v(y,t) = yw + Ew\sum_{k=1}^{t-1} A_k - s - (t-1)g_0 \tag{4.4.17}$$

This simplified version of the dynamic programming rule will be called SDPR. The profit of the SDPR may be a little smaller than the profit of the DPR, but the choices for each state and number of new orders can be found much easier. Again we can start for instance with $g_0 = g_\pi$ and determine the choices $Ch(y,t,j)$ in the same way as in the DPR, although now the values of $v(y,t)$ are not directly influenced by this choice. From this set of choices we can determine the average costs of the rule in exactly the same way as in the DPR by using (4.4.7)-(4.4.15) and then use this value g_0 in (4.4.17) and repeat this procedure again.

As we already suggested, it is not necessary to consider all the possible choices in a particular state, with a particular demand. Therefore we can use the following properties.

Property 4.3 : for a given demand j the proposed lead time in state (y_1,t), $Ch(y_1,t,j)$ will be no longer than the proposed lead time in the state (y_2,t),$Ch(y_2,t,j)$, with $y_1 < y_2$.

Property 4.4 : for a given state (y,t) the proposed lead time for demand j_1, $Ch(y,t,j_1)$ will be no longer than the proposed lead time for demand j_2, $Ch(y,t,j_2)$, with $j_1 > j_2$.

Proof property 4.3: For given values of the state y,t and the demand j, the lead time $Ch(y,t,j)$ is determined in formula (4.4.4) by finding the k for which the function

$W(j,y,k)$ is maximised. This function $W(j,y,k)$ is the same as $dc(y,t,j,Ch)+\sum_{x,k} Q(y,t;x,k)\cdot v(x,k)$ in (4.4.4) and is defined as

$$W(j,y,k)=P^k_{0j}v(y,t-1)+\sum_{i=1}^{j}P^k_{ij}(v(y+i,k)-hy(t-1-k)) \tag{4.4.18}$$

In the SDPR, $v(y,t)$ is given by (4.4.17). Suppose that for given values of y,t and j the proposed lead time is given by $Ch(y,t,j)=k$, with $2\le k\le t-1$. This implies that for this state and demand

$$W(j,y,k)\ge W(j,y,k_1) \qquad 1\le k_1<k \tag{4.4.19}$$

Using (4.4.17), the difference $W(j,y,k)-W(j,y,k_1)$ can be written as

$$W(j,y,k)-W(j,y,k_1)=jw(A_k-A_{k_1})+Ew\sum_{l=k_1}^{k-1}A_l-(k-k_1)(g_0-hy)$$

$$+P^k_{0j}(Ew\sum_{l=k}^{t-2}A_l-(t-1-k)(g_0-hy))$$

$$-P^{k_1}_{0j}(Ew\sum_{l=k_1}^{t-2}A_l-(t-1-k_1)(g_0-hy)) \tag{4.4.20}$$

In the state $(y+n,t)$, with $n\in N$, we will prefer a lead time k above a lead time $k_1<k$ if $W(j,y+n,k)\ge W(j,y+n,k_1)$. Replacing y by $y+n$ in (4.4.20) yields:

$$W(j,y+n,k)-W(j,y+n,k_1)=W(j,y,k)-W(j,y,k_1)-hn(k-k_1)$$

$$+P^k_{0j}(t-1-k)hn-P^{k_1}_{0j}(t-1-k_1)hn$$

$$=W(j,y,k)-W(j,y,k_1)+$$

$$+hn(k-k_1)(1-P^{k_1}_{0j})+$$

$$+(P^k_{0j}-P^{k_1}_{0j})(t-1-k)hn$$

Since all terms in the last equation are non-negative or positive, we have that for the state $(y+n,t)$ and the demand j

$$W(j,y+n,k)-W(j,y+n,k_1)>0$$

which completes the proof of property 4.3. □

Proof property 4.4: In this proof we use the same function $W(j,y,k)$ as defined in (4.4.18), but now we are interested in the best lead time k for different values of the demand. Suppose that for given values of y,t and j the proposed lead time is given by $Ch(y,t,j)=k$, with $1\le k\le t-2$. This implies that for this state and demand

$$W(j,y,k) \geq W(j,y,k+1)$$

Using (4.4.20), the difference $W(j,y,k)-W(j,y,k+1)$ can be written as

$$W(j,y,k)-W(j,y,k+1) = jw(A_k-A_{k+1})-EwA_k+g_0-hy$$

$$+P_{0j}^{k}(Ew\sum_{l=k}^{t-2}A_l-(t-1-k)(g_0-hy))$$

$$-P_{0j}^{k+1}(Ew\sum_{l=k}^{t-2}+1\,A_l-(t-1-k+1)(g_0-hy))$$

which we rewrite, for notational reasons, as

$$=jC_0-C_1+x^j(C_1+C_2)-y^jC_2$$

where all terms are non-negative and $0<x<y<1$. We also define

$$VW(j)=W(j,y,k)-W(j,y,k+1)-W(j-1,y,k)+W(j-1,y,k+1)$$

$$=C_0+(x-1)x^{j-1}(C_1+C_2)-(y-1)y^{j-1}C_2$$

Now if

$$k=Ch(y,t,l)<Ch(y,t,l-1) \tag{4.4.21}$$

for some $l \in N$, then $VW(l)>0$, implying that

$$C_0>(1-x)x^{l-1}(C_1+C_2)-(1-y)y^{l-1}C_2$$

Multiplying the right-hand side by $y^m, m \in N$, we have

$$C_0>(1-x)x^{l+m-1}(C_1+C_2)\left[\frac{y^m}{x^m}\right]-(1-y)y^{l+m-1}C_2$$

and since $y>x$, we have

$$C_0>(1-x)x^{l+m-1}(C_1+C_2)-(1-y)y^{l+m-1}C_2$$

$$=C_0-VW(l+m)$$

which implies that

$$VW(l+m)>0 \qquad \textit{for all } m \in N \tag{4.4.22}$$

and because of the definition of $VW(.)$ and because of (4.4.21), we have that

$$W(l+m,y,k)-W(l+m,y,k+1)>0$$

and thus $Ch(y,t,l+m) \leq k=Ch(y,t,l)$ which completes the proof of property 4.4. □

By using these two properties we can avoid much of the work for determining the values of $Ch(y,t,j)$. In the next subsection we will compare this rule with the DPR and with the cyclic rule π.

4.4.6. Numerical results

In this subsection we want to compare the rules we discussed for the generation of the lead times. Therefore, we consider several examples and compare the average profit, the average storage costs, the percentage of withdrawals and the average time between set-ups (TBSU). We will consider a situation with set-up costs $s = 100$, holding costs $h = 1$, revenues per item ordered $w = 40$ and $A_i = 1.1-0.1\,i$; $i = 1,..,10$ and 0 elsewhere. In all examples we will have $E = 2$, but the variance of the demand will be different in the examples. For different values of M, the demand will be distributed as: $d_0 = (M-3)/(M+1)$ and $d_j = 4/(M(M+1))$ for $j=1,..,M$. The values of M that will be considered are 5,8,10,12 and 15. The results for the rules will be given in the Tables 4.4,4.5 and 4.6.

M-value	profit	TBSU	max. N	% withdr.
5	44.140	5.035	5	20.00
8	45.574	4.459	4	15.00
10	47.435	4.071	3	10.00
12	50.063	3.855	2	5.00
15	55.000	4.000	1	0.00

Table 4.4. the results for the cyclic rule π

In the rule π we have no storage costs, so these costs are not given in the Table 4.4. The maximum lead time N is always 5 when we use the DPR and the SDPR, so the N-value is not given in the last two tables. Comparing the three rules we see that the SDPR and the DPR are only different for large M. The reason for this is that for smaller M the values of $v(y,t)$ are nearly the same in both rules, because most of the orders arriving in state (y,t) will obtain a promised lead time that equals $t-1$. If the number of orders is large, a smaller lead time can be proposed, but this will only be

M-value	profit	TBSU	storage	% withdr.
5	45.174	4.770	0.007	17.32
8	47.801	4.707	0.037	13.64
10	50.439	4.243	0.059	7.41
12	53.309	4.322	0.050	4.38
15	57.075	4.687	0.024	1.96

Table 4.5. the results for the SDPR

M-value	profit	TBSU	storage	% withdr.
5	45.174	4.770	0.007	17.32
8	47.801	4.707	0.037	13.64
10	50.514	4.308	0.070	7.76
12	53.379	4.367	0.062	4.58
15	57.186	4.839	0.046	2.63

Table 4.6. the results for the DPR

done for larger values of M. There is a clear difference between the cyclic rule and the other two strategies with respect to the profit but the patterns for the profit and for the percentage of withdrawals show a lot of resemblance.

Chapter 5

SEVERAL TYPES OF PRODUCTS ON ONE MACHINE

5.1. Introduction

In Chapter 4 we have observed that in a situation with no capacity restrictions a rather simple production rule can be used, the (x,T)–rule. In the (x,T)–rule the dimension of the state space is largely reduced before the production decision is made. This reduction allows a simple production rule. We are interested whether in a more complex situation such a type of rule can also be used. In this chapter and in the next chapter, we will consider a situation in which several types of products are produced on one machine. The amount of work that can be done on the machine is restricted, due to the machine speed and to the time the machine is available for production. If the production is changed from one type of product to another, a set-up is needed. By performing a set-up, capacity is lost. Loosing too much capacity due to set-ups will lead to large delivery times for the ordered goods, but on the other hand, waiting too long with a set-up will also lead to large delivery times. By controlling the production, a balance has to be found such that the clients can be satisfied by limiting the number of late orders or by limiting the lateness of the orders, while on the other hand also the manufacturer can be satisfied by limiting the set-up costs, the holding costs and in some situations also the costs for working overtime.

In Section 4.2. we have studied a situation with fixed lead times and with no capacity restrictions. The problem for the manufacturer was to produce the orders such that the requested lead times were realised with limited costs and in a way that was satisfactory for the customers. In this chapter we will consider an extension of this situation. The orders will again have fixed lead times, but now there are capacity restrictions. Therefore we can now study the situation in which the customers may order products of several different types. This extension makes the production rules more complex because now several types compete for the same capacity. We will consider two possibilities for the available capacity. The available capacity can be fixed or there may be possibilities for working overtime or other ways of flexible capacity. We will model this situation as a multi-type capacitated problem with periodic review.

5.1.1. Description of the model

In the model we assume that M types of products can be produced on the same machine. The length of a period is C time units and we will express the delivery times and the lead times in an integer number of periods. The maximum lead time will be N periods. The service time, that is the time to produce one order, is supposed to be one time unit for all types. Every set-up takes S time units. Producing a type of product implies a set-up, disregarding the possibility that the same type may have been produced in the previous period. This assumption is made because this element is quite common in periodic review models, for instance in the uncapacitated Wagner-Whitin approach. The assumption has also a large influence upon the production rules, but the rules that will be presented can easily be adapted to a situation in which no set-up would be needed if production continues in the next period. By these assumptions, the maximum number of orders that can be produced in one period, without working overtime, is equal to $C-S$.

We assume a gating service discipline, implying that only those orders can be produced, that were ordered before the production period. In the situation in which extra capacity is available, this extra capacity will only be used for the production of goods that were ordered before the start of the production period. The demand is supposed to be stationary and independent of the schedule and of the delivery times that have been realised previously. Although it is not very essential for the production rules, we will assume that the orders will be placed by $N \times M$ groups of clients, where the demand from different groups of clients is assumed to be independent, with lead times varying from 1 to N.

At the end of a period we know the exact required deliveries for the next period, this period will be called the first period, and we know a part of the required deliveries for later periods. Apart from this order state vector, we also know the distribution of the future demand. Based on this knowledge, we can decide to produce the required deliveries for one or more types of products, or decide to delay the production at least one period. The direct costs of such an action are about the same as in Section 4.2. Every set-up takes set-up costs s. In those periods in which we produce, the total amount of capacity that is needed is the sum of the set-up times S and the number of orders we want to produce. If this amount exceeds C and if we can do extra work then we have to pay overtime costs of z per unit of extra capacity. For orders that are manufactured before the due date period we have holding costs h per order per period and for orders that are delivered too late we have penalty costs p per order per period. Late orders are not lost, but they are backlogged.

In Section 5.2. we will assume that no extra capacity is available, whereas in Section 5.3. extra work can be done, involving extra costs. We will consider the same production rules for situations in 5.2. and 5.3., since the possibility of using extra capacity will be the only difference between the two situations. For these situations we want to use simple production rules. We will start with an extension of the (x,T)-rule. Then we will give a production rule that is inspired by well-known production rules for multi-type capacitated lot sizing problems. We will also consider a cyclic production rule. The performance of the production rules will be measured by the average costs per period. These average costs will be compared in a simulation study in which we will also consider the computation time that is needed for use of the different production rules. Before we describe the rules, we will start with some notations.

5.1.2. Notation

In an arbitrary period t, goods may be ordered for the periods $t+1, t+2, .., t+N$. For each type of product and for each of the N periods the distribution of this demand is independent of t and of any of the other demands for other types or other periods. Now we use the following notations with $1 \le i \le M$, $1 \le j \le N$:

- r_{ij} is the order state for products of type i for period j. The order state contains all unfinished orders for this period. Because backlogging is permitted, r_{i1} also contains orders from earlier periods which are not yet produced. We write R for the $M \times N$ matrix containing the complete order state.

- $m_{ij} = \sum_{k=1}^{j} r_{ik}$ is the cumulative order state for products of type i up to and including period j.

- e_{ij} is the expected value of the j-th component of the order state vector for products of type i, if no jobs ordered for this period and type have been produced,(cf. 4.2.4.1).

- a_i is the action we take for type i during the first period. The meaning of action a_i is that we produce a_i orders of type i. If there is no extra capacity available, a_i cannot be larger than $C-S$. Action 0 means that we do not produce. We write A for the vector containing the actions for all types of products.

- $q(r_i, a_i)$ is the one-stage cost of taking action a_i on having the order state vector r_i for type i and has the following form:

$$q(r_i, a_i) = p\,(r_{i1} - a_i)^+ + s\,(1 - \delta(a_i)) + h \sum_{j=1}^{N-1} (a_i - m_{ij}) 1_{(0,\infty)}(a_i - m_{ij}).$$

Here $\delta(a_i) = 1$ for $a_i = 0$ and 0 elsewhere, $(x)^+ = x$ if $x > 0$ and 0 if $x \le 0$, and $1_{(0,\infty)}(x)$ equals 1 if $x \in (0,\infty)$ and 0 otherwise. The total one-stage costs of taking the actions according to vector A for the complete order state R in the situation in which there are no possibilities for working overtime, have the following form:

$$q(R, A) = \sum_{i=1}^{M} q(r_i, a_i)$$

In the situation in which extra work can be done, the total one-stage costs are given by:

$$q(R,A) = \sum_{i=1}^{M} q(r_i,a_i) + z\,(\sum_{i=1}^{M} a_i + S \sum_{i=1}^{M} (1-\delta(a_i)) - C)^+$$

5.2. No possibilities for working overtime

5.2.1 Introduction

In this section we will consider the dynamic multi-type, capacitated problem with no possibilities for working overtime. The objective of the manufacturing firm is to minimise the long term average costs per period. Since finding the optimal policy will be very hard for large problems, because of the enormous number of possible order states, we will concentrate our attention on finding a good heuristic, that is a heuristic with low average costs which is also very fast to use.

Next to the cyclic production rule, the (x,T)–rule has proven to be such a simple and fast rule in the uncapacitated situation. The (x,T)–rule in the form in which it has been described in 4.2.4., should be adapted a bit to the extra problems of a restricted capacity and the variety of types competing for the same capacity on one machine. We will deal with these problems in a simple way and the resulting production rule will be called the extended (x,T)–rule. In order to judge the performance of this extended (x,T)–rule, we want to compare it with a more complex production rule which contains several elements of well-known production rules.

The static version of this problem, where the demand is known completely for a number of periods, the so-called capacitated, multi-type, time-varying demand problem, has been one of the most favourite topics in the area of production scheduling and a number of heuristics have been developed to solve this problem (see, for example, Dixon and Silver (1981), Dogramaci et al. (1981), Florian and Klein (1971) and Lambrecht and Vanderveken (1979) and Maes and Van Wassenhove (1986)). Although Dixon and Silver also use a two-step approach, the procedure presented by Van Nunen and Wessels (1978) has been the most inspiring heuristic for the second production rule we will consider for this problem, the so-called two-step rule. The Van Nunen and Wessels approach is to first ignore the capacity restrictions and solve the problem for each individual type independently, using an extension of the Wagner and Whitin method like we have done in Section 4.2.. The resulting overall solution is usually unfeasible, violating the capacity restrictions in one or more of the production periods. The next step is to adjust the solution until it is feasible, trying to make the increase in the costs as small as possible.

The fixed cycle production rule can be used directly in the multi-type capacitated situation. However, in this rule we assume that we have a production interval of a fixed length for every type of product. Since the length of a period is also fixed, this may lead to problems. Usually it makes sense to produce more than one type of product in one period. This cyclic rule in which we can produce different types of products during one period with a given length for the period will be called a semi-fixed cycle rule. The performance of this rule, the extended (x,T)–rule and the two-step approach will be compared in a few examples at the end of this subsection.

5.2.2. Extended (x,T)-rule

5.2.2.1. Introduction to the extended (x,T)-rule

In this section we will describe the adaptation of the (x,T)–rule in the form in which it has been described 4.2.4. to the extra problems of a restricted capacity and the variety of types competing for the same capacity on one machine. We would like the resulting production rule, which will be called the extended (x,T)–rule, to be a simple and fast heuristic, like the original (x,T)–rule. We will adapt the uncapacitated (x,T)–rule to the capacity restrictions by using *penalty points*.These penalty points indicate how urgent the production of an order or the production of a type is. Penalty points can be allocated to orders in every thinkable way, it may be a function depending on the type, on the arrival period, on the due date and so on. In general, the penalty points for a type will be the sum of the penalty points of all orders that type, but an additional element can be added.

The idea behind the production rule is now, to produce the known orders that should be delivered within a certain number of periods, T say, if the number of penalty points of the type is large enough, for instance, at least x. If several types are competing for the same capacity, capacity is allocated to the types according to a decreasing number of penalty points. If capacity has been allocated to one or more types and the remaining available capacity is too small to make an allocation to one of the other types profitable, we can try to produce some extra orders for one of the types, to which capacity has been allocated. In this way, it is possible to produce orders with a residual lead time which is more than the usual maximum of T periods ahead, thus avoiding possible penalty costs due to future capacity shortages.

5.2.2.2. Formulation of the extended (x,T)-rule

In order to determine good values for the penalty points and for the pair (x,T), we can first study the uncapacitated situation. For instance, we can determine the optimal pair (x_i,T_i) for every type of product $i=1,2,..,M$. In the uncapacitated situation of Section 4.2.4.4. we have chosen the penalty points to be 1 for the orders that should be produced by the end of the period, the component r_1, and 0 for all other orders. An alternative is to choose the penalty points to be p for the orders that should be produced by the end of the period and 0 for all other orders and to choose the minimum level for production x_i equal to $g_i(x_i,T_i)$, being the average costs of the uncapacitated (x,T)–rule for type i. According to property 4.1. this alternative will usually be the same as the original (x_i,T_i)-rule. In the multi-type capacitated situation the choice of x and its dependence on the penalty costs p differs somewhat from the uncapacitated situation;

1) Because of the competition with the other types, there is not always capacity available for a type for which the number of penalty points would justify a set-up. Therefore, it will be wise to choose the penalty points and the level x_i in such a way that, if they were chosen the same in the uncapacitated situation, they would lead to a more frequent production.

2) In the situation with the restricted capacity and set-up times, future shortages can be avoided best by producing large lots for a few types. Therefore we give penalty points to *all* orders of which production seems profitable from the point of view of total costs and not only to the orders that should be delivered before the end of the period. The effect of this is more or less the same as in a production rule where we select the types on the costs per produced order. Analogous to the previous rule, the penalty points can be seen as the difference between on the one hand the direct and future penalty costs reduction and on the other hand the extra holding costs. Of course, the penalty points for orders that should be delivered before the end of the period, should be much higher than for

orders with due dates in later periods.

3) In the competition for the available capacity, we prefer the type(s) with the highest number of penalty points. To avoid discrimination of the types with a small average demand, the penalty points for these types should be chosen larger than those of the types with a high average demand. Maes and Van Wassenhove (1986) suggest several criteria, such as time-between-set-ups, to set priorities for the different types. Graves (1980) uses the demand rate as a scaling factor. It is often impossible to determine the best choice for such a scaling factor, but in general a scaling factor will be used which is between 1 (no scaling) and the demand rate (equal time-between-set-ups).

The elements sketched above will give an indication for the choice of the penalty points and the required minimum x_i. Finding a good choice will now be a question of trial-and-error combined with some intuition. Now we will describe the extended (x,T)–rule for a given set of penalty points, with f_{ij}, $i=1,..,M$, $j=1,..,N$, the penalty points for an order of type i with a residual lead time of j periods. We also assume that the values for T_i, $i=1,2,..,M$ are determined according to the uncapacitated (x,T)–rule according to the method described in 4.2.4.4.

The penalty points for a type, written as $K(i)$, are defined by:

$$K(i)=\sum_{j=1}^{N} f_{ij} r_{ij} \tag{5.2.1}$$

If $K(i)<x_i$ for all i, then we do not produce at all during the period. Otherwise, we determine the sequence of types $\{i_1,i_2,..,i_k\}$, with $k \leq M$, in which the types with a sufficient amount of penalty points are placed according to a decreasing amount of penalty points, $K(i_1) \geq K(i_2) \geq .. \geq K(i_k)$. Types with an equal number of penalty points are placed according to an increasing average demand. In the uncapacitated situation we would produce all the types in this sequence and the amount for type i_l would be:

$$a_{i_l}=\sum_{j=1}^{T_{i_l}} r_{i_l j} \tag{5.2.2}$$

In the multi-type capacitated situation the actions may be different. We start with a capacity allocation of $\max(C, a_{i_1}+S)$ for type i_1 leading to a remaining capacity $C_1=C-a_{i_1}-S$ or $C_1=0$. If the remaining capacity is less than a certain minimum capacity level *mcl*, the remaining capacity will not be spent on a set-up and an allocation for the following type in the sequence. The idea is that in the situation in which the remaining capacity is small, it will be better to use the capacity for production than for a set-up. If $C_1 \geq mcl$ then we continue with the second type i_2 and make an allocation of $\max(C_1, a_{i_2} + S)$ for type i_2 leading to a remaining capacity C_2.

In this way we continue until either an allocation has been made for all types in the sequence, which gives the combination of uncapacitated actions, or until the remaining capacity is less than *mcl*. If the remaining capacity after the first set of capacity allocations is larger than 0 and less than the extended capacity level, denoted by *ecl*, with $ecl \geq mcl$, we start a second round of the capacity allocations. *ecl* and *mcl* are both decision variables, which can be chosen freely. In this second allocation round we will consider the production of orders with a residual lead time of one period after the horizon T_i. This will be done only for the types, for which a capacity allocation has been made, following a sequence based on an increasing average demand. If there is still remaining capacity after the second round or if the remaining capacity after the first round is at least *ecl*, then we do not use the remaining capacity.

5.2.3. The two-step rule

5.2.3.1. Introduction to the two-step rule

Following the Van Nunen-Wessels approach, we split the rule in two parts: a part for the individual types and a part on the aggregated level. In the first part we determine the expected marginal costs of all possible actions for each type separately, thereby assuming capacity restrictions for the first period only. We choose a good combination of actions based on the sum of the separate costs, the so-called *base costs*. In the second part, which is only necessary if there is still capacity available after the first part, we consider the possibility of producing extra orders to avoid the future costs due to capacity shortages. These future costs are calculated on a higher aggregation level, thus avoiding much computational work.

5.2.3.2. Formulation of the first part

In the first part of the rule we determine the marginal costs for the possible actions for each of the types. Therefore we use the Wagner-Whitin like approach for the individual types, as described in 4.2.4.3. This rule uses the expected total costs as its only criterion for determining the best action, unlike for instance the (x,T)–rule and the Silver-Meal approach. This is a big advantage, since we want to compare actions for different types and in the second part also costs on an aggregated level. We will only consider the first periods action. Therefore the Wagner-Whitin approach will be used, including a salvage function, with a horizon of 1 period. Of course, we can produce orders with a residual lead time of more than 1 period.

To measure the effect of the first period action upon the costs during the following periods, we will use a salvage function, as introduced in Section 4.2.4.3. As an important element in the salvage function we determine the average costs per period, g_π^i, of the best cyclic production rule π_i, for every type i, i=1,2,..,M. The average costs can be found by determining the best value for T_i, the time between set-ups, in formula (4.2.10). Now the marginal costs of an action a_i for type i can be described as the sum of the direct costs, $q(r_i,a_i)$, and the salvage function $L_i(a_i)$.

Let $z_{ij}^k(a_i)$ be the expected value for the j-th component of the order state vector for product type i at the beginning of the k-th period, if in the first period action a_i has been taken for type i and if in later periods no production has taken place, (cf. (4.2.1) and (4.2.11)). Let $H(a_i)$ be the first period for which the penalty costs for type would be larger than g_{π_i}:

$$H(a_i) = \min\left\{ k=2,3,.. \middle| p \cdot z_{i1}^k(a_i) > g_{\pi_i} \right\}$$

The salvage function is now defined as the expected sum of the costs during the periods 2,..,$H(a_i)$ reduced with $H(a_i)$ times g_{π_i}, under the following assumptions:

1) During the first period a_i orders are produced for type i.

2) If $H(a_i)$>2 we do not produce during the periods 2,..,$H(a_i)-1$.

3) In period $H(a_i)$ we produce the required deliveries for the first T_i periods.

The salvage function $L_i(a_i)$ thus defined, is slightly different from the salvage function $L(A)$ defined in (4.2.12) and (4.2.13). Now, the salvage function for type i is given by:

$$L_i(a_i) = \sum_{k=2}^{H(a_i)-1} p\, z_{i1}^{k}(a_i) + s + h \sum_{j=2}^{T_i} (j-1) z_{ij}^{H(a_i)}(a_i) - H(a_i) g_{\pi}^{i} \tag{5.2.3}$$

In the first step of the rule we start with the calculation of the sum of the direct costs and the salvage function, the so-called *base costs*, for every type and every possible action.

Determine $Co(i,a_i) = q(r_i,a_i) + L_i(a_i)$ for $i = 1,..,M$, $a_i = 0,..,C-S$.

We define $Co(i,j) = \infty$ if $a_i > m_{iN}$, that is, if there are not enough orders for type i to make action a_i possible. In the situation we consider, it is possible to produce more than one type during a period. Therefore it is better to consider the total use of capacity instead of an action. For action a_i the use of capacity is a_i+S. Now we define the *base cost reduction* $K_0(i,j)$ as the difference in base costs between the use of j units, i.e. action $a_i = j-S$, and the use of 0 units of capacity for type i. Here the subscript '0' indicates the number of types to which capacity has been allocated.

Determine $K_0(i,j) = Co(i,j-S) - Co(i,0)$ for $i=1,..,M$, $j=S,..,C$.

We define $K_0(i,j) = s$ if $0 < j \le S$ and $K_0(i,0) = 0$ for $i=1,..,M$. Now the capacity 'action' for which the base cost reduction $K_0(i,j)$ is minimal is the best 'action', at least as long as we restrict ourselves to the production of one type.

Determine the pair (i,j) for which $K_0(i,j)$ *is minimal with* $i=1,..,M$, $j=1,..,C$.

Assume that the base cost reduction is minimal for the pair (i_0,j_0). If it is optimal to use no capacity, that is if the minimum value is larger than 0, then we continue with the second part of the rule. Otherwise, j_0 units of capacity are allocated for a set-up and the production of j_0-S orders for type i_0. If all capacity is used in this way, that is if $j_0=C$, then we are finished for this period. If not, we determine the remaining capacity $C_0 = C - j_0$ and we adapt the base cost reduction for the type i_0:

Determine $K_1(i_0,j) = K_0(i_0,j+j_0) - K_0(i_0,j_0)$ for $j=0,..,C_0$.

Notice that the values $K_1(i_0,j)$ are all non-negative, so a type can be chosen only once in this first part of the rule. However, these values can be important in the second part of the rule. For the other types the base cost reduction $K_1(i,j)$ is not different from $K_0(i,j)$. Now the production of orders for one of the other types will be considered.

Determine the pair (i,j) for which $K_1(i,j)$ is minimal with $i=1,..,M$, $j=1,...,C_0$.

Let the base cost reduction be minimal for the pair (i_1,j_1). Now we can make the same decisions as for the previous pair: continue with the second part if the minimal value is larger than 0, stop if j_1 equals C_0 or continue with the first part after adapting the capacity values for the type i_1 and the remaining capacity $C_1=C_0-j_1$. Finally we come to a situation, in which either all available capacity is used, or in which the base cost cannot be reduced any more.

5.2.3.3. Formulation of the second part

If the available capacity is not allocated completely in the first part of the rule we continue with the second part. In this second part we shall consider the possibility of producing some extra orders to avoid capacity problems in the future and we determine the effect of this production on the costs on the aggregated level. Calculating this effect exactly is usually impossible, because of the complex state and action spaces. Therefore the first step is to simplify the state and action space by means of aggregation. The order state on the aggregated level takes the form:

$$(r_{a1},r_{a2},...,r_{aN}) := (\sum_{i=1}^{M} r_{i1},\sum_{i=1}^{M} r_{i2},...,\sum_{i=1}^{M} r_{iN})$$

and the action on the aggregated level is defined as

$$a_a := \sum_{i=1}^{M} a_i$$

Let e_{ai} be the expected number of orders on the aggregated level that arrive during the last i periods before their due date period t:

$$e_{ai} = \sum_{j=1}^{M} e_{j1} - \sum_{j=1}^{M} e_{ji+1} \qquad i=1,2,..,N-1 \qquad (5.2.4)$$

$$e_{aN} = \sum_{j=1}^{M} e_{j1} \qquad (5.2.5)$$

Then we can describe the expected number of orders that should be produced before the end of the i-th period, denoted by F_i, in the following way:

$$F_2 = r_{a1} + r_{a2} + e_{a1} \tag{5.2.6}$$

$$F_i = F_{i-1} + r_{ai} + e_{ai-1} \qquad i=3,..,N \tag{5.2.7}$$

$$F_i = F_{i-1} + e_{aN} \qquad i>N \tag{5.2.8}$$

The maximum number of orders that can be produced during a period is $C-S$. So, if $r_{a1}+r_{a2}>a_a+(C-S)$, we are certain of having penalty costs due to capacity shortages in the second period. In order to keep the production rule quite simple and fast, we want to avoid that the aggregated action is an N-dimensional vector. Therefore we assume that the aggregated action influences F_j for all $j\geq 2$, of course allowing $a_a>r_{a1}+r_{a2}$. Due to this assumption, increasing a_a by 1 unit, decreases the expected penalty costs in the second period by p units. However, in many situations the effect of increasing a_a will not be as clear as in this situation. A reason for this is that the maximum value of a_a depends on the number of types that will be produced. If in the cyclic rules π_i we have an average of u set-ups per period, where u is possibly rounded to the nearest integer, it is more reasonable to assume that the maximum action on the aggregated level is $C_{max}=C-uS$. It may even be so that by producing some extra orders of a type, the penalty costs increase, due to a delay of the next production of this type for one or more periods. Therefore, to our belief, it makes no sense, to calculate the expected penalty costs due to capacity shortages in a very accurate way, by considering the demand distribution for the future periods and considering the probabilities on certain losses of capacity due to more than one set-up in a period. Instead of this accurate estimation, we advocate a much more simple estimation, which can be performed for example in one of the following three ways:

1) We do not consider any variability, but we just consider the penalty costs we would have if the number of orders would equal F_i and if, from the second period on, we would use all available capacity, by taking the maximum action C_{max}. The penalty cost function in this model is given by:

$$P_1(a_a) = p\sum_{i=2}^{H}\left[F_i-(i-1)C_{max}-a_a\right]^+ \tag{5.2.9}$$

In the summation we can stop after H periods, where H can be chosen freely, or as soon as we have a non-positive term for $i>N$.

2) In the second method of estimating future penalty costs we will consider the original variability for the cumulative number of orders in future periods. We estimate the penalty costs by assuming that the number of orders is distributed

according to a uniform distribution and that every period the maximum action C_{max} is taken. We have to determine the lower limit and the upper limit for the cumulative number of orders for 'every' period and compare this with the sum of the available capacity and a_a. By using a positive multiplying constant *mc* for the deviation of the number of orders we can influence the effect of the second part of the rule. The best value for this constant depends on the tightness of the capacity and on the values for *p*,*h* and *s*. Writing v_{ai} for the known variance, x_{ai} for the maximum deviation, l_{ai} for the lower limit during period *i* and u_{ai} for the upper limit, and assuming the uniform distribution, we have:

$$x_{ai} = mc\sqrt{(12\sum_{j=2}^{i} v_{aj})} \tag{5.2.10}$$

$$l_{ai} = F_i - x_{ai} \tag{5.2.11}$$

$$u_{ai} = F_i + x_{ai} \tag{5.2.12}$$

For every period there are three possibilities for the penalty cost function: if the sum of the available capacity and the aggregated action exceeds the upper limit for the cumulative number of orders, there are no penalty costs, if the sum is between the lower and the upper limit, there are partial penalty costs and if the sum is less than the lower limit, the penalty costs are complete, that is like in (5.2.9):

$$P_2(a_a) = p\sum_{i=2}^{H}\Big((F_i-(i-1)C_{max}-a_a)1_{(0,l_{ai}]}((i-1)C_{max}+a_a)+$$

$$+\frac{(u_{ai}-(i-1)C_{max}-a_a)^2 1_{(l_{ai},u_{ai}]}((i-1)C_{max}+a_a)}{4x_{ai}}\Big) \tag{5.2.13}$$

We can stop the summation as soon as the sum of the available capacity and the aggregated action exceeds the upper limit, $(i-1)C_{max}+a_a > u_{ai}$ and $i>N$ or after a fixed number of periods *H*.

3) In this third method we do not consider any variability in the demand, but instead we assume that the value of the penalty cost function decreases with every order that is produced. For every period we determine the expected number of orders. As long as the expected number of orders is larger than the available capacity minus set-up times, a multiple of C_{max}, the penalty function decreases with an amount *p*. If the expected number of orders is *k* units smaller than the multiple of C_{max}, then the penalty function decreases with an amount $p\alpha^k$, where α is a

constant that can be chosen freely between 0 and 1. The choice for the value of α that yields the smallest average costs depends on the tightness of the capacity and on the the values for s, h and p. This results in the following penalty function:

$$P_3(a_a)=p\sum_{i=2,..,H}\Big((F_i-((i-1)C_{\max}+a_a)+\frac{1}{1-\alpha})1_{(0,F_i]}((i-1)C_{\max}+a_a)+$$
$$+\frac{\alpha^{((i-1)C_{\max}+a_a-F_i)}}{1-\alpha}1_{(F_i,\infty)}((i-1)C_{\max}+a_a)\Big) \qquad (5.2.14)$$

In the summation we can stop after H periods, where H can be chosen freely, or as soon as we have a very small term for $i>N$.

Now we return to the description of the algorithm. If in the first part available capacity is not allocated entirely, we come to this second part to consider the effect of the aggregated action a_a. Because we have taken already some decisions about the production in the first part, we do not have to consider all possible values for a_a. If in the first part capacity has been allocated to k different types and the remaining available capacity is C_{k-1}, then the lower value we have to consider for a_a, written as $a_{\min}$ is the sum of the actions taken in the first part:

$$a_{\min}=\sum_{i=0}^{k-1}(j_i-S) \qquad (5.2.15)$$

and the upper value, $a_{\max}$ is determined by adding the remaining capacity to $a_{\min}$:

$$a_{\max}=a_{\min}+C_{k-1} \qquad (5.2.16)$$

For these values of a_a, we determine the penalty cost value $P_1(a_a)$, or as an alternative, $P_2(a_a)$ or $P_3(a_a)$. Since we are merely interested in the differences in penalty cost value between different values of a_a, we define the marginal penalty costs, written as $P^*(j)$.

Determine $P^*(j)=P_1(a_{min})-P_1(j+a_{min})$ *for* $j=0,1,..C_{k-1}$.

This non-negative function $P^*(j)$ gives the reduction in the penalty cost function value if we take the aggregate action $j+a_{\min}$ instead of $a_{\min}$. The remaining part of the rule is quite easy. If the minimum value of the extra base costs minus the marginal penalty costs, $K_k(i_k,j_k)-P^*(j_k)$, is positive, then we stop. If the minimum is zero or negative, then the best 'action' (i_k,j_k) will be taken, which implies that j_k units of capacity are allocated to type i_k. Then we adapt the remaining capacity, $C_k=C_{k-1}-j_k$, the values for the base costs for the type i_k and the marginal penalty costs $P^*(j)$. If there is no more remaining capacity, we stop, but otherwise we continue with the next pair.

Determine the pair (i,j) *for which* $K_{k+1}(i,j)$ *is minimal with* $i=1,..,M$, $j=1,..,C_k$.

Finally, if the extra base costs are larger than the penalty cost reduction or if all available capacity is used, the planned actions can be performed and the procedure can be repeated the next period.

In Section 5.2.5. some examples are given. From these examples we will learn the performance of the penalty cost functions for different parameter values and we will also consider the CPU-time for the simulation with the three penalty cost functions.

5.2.4. A semi-fixed cycle rule

5.2.4.1. Introduction

A production rule that is even more simple than the extended (x,T)–rule is the fixed cycle production rule. As described in Chapter 3, a fixed production cycle is a cycle in which both the sequence in which the types will be produced as well as the available capacity for the production of a type is fixed. The fixed cycle rule follows this production cycle, which is repeated over and over again. In the situation with a normal gating service discipline without possibilities for working overtime, which we are considering now, the fixed cycle rule assumes that during a period only one type of product can be produced. Sometimes this is not a problem, for instance if we find that the best value for the time between set-ups, T_i in the cyclic production rule π, which can be found by formula (4.2.10), is for all types equal to M periods, where M is the number of different types. In that situation it is very logical that in a fixed cycle we produce exactly one type every period.

If we find however, that for even M the best value for the time between set-ups, T_i, is equal to $\frac{1}{2}M$ for every type i, then it is not logical to produce exactly one type every period. In that situation, we have two options: we split up the original period in two equal parts, which become the new periods, or we decide to produce two different types in one period, dividing the available capacity according to the (varying!) capacity requirements for both types. A rule in which the production takes place according to the second option, that is with a cycle in which we can produce a number of different types during one period with a given length for the period, will be called a semi-fixed cycle rule . We can consider the fixed cycle production rule as a special case of the semi-fixed cycle rule, with only one type of product every period.

5.2.4.2. Determining a semi-fixed cycle

If we want to determine the production cycle in this multi-type capacitated situation, two elements are important: the (maximum) number of types we can produce during a period, the so-called *multi–type level*, and the relative frequency of a type. If we denote the multi-type level by L and the relative frequency for type i by rf_i, then the absolute frequency of type i, denoted by af_i, is given by:

$$af_i = L\frac{rf_i}{\sum_{j=1}^{M} rf_i}$$

The absolute frequency of a type represents the average number of production opportunities for a type during one period. The two important elements, the multi-type level and the relative frequency, can both be chosen freely.

We can determine the multi-type level in a simple way by calculating T_i, the average time between set-ups for type i, using for instance the cyclic production rule π_i or the uncapacitated (x,T)–rule. From these T-values we can determine the average number of set-ups. Then we set the multi-type level equal to the rounded value of the average number of set-ups. Of course we have to verify that the available capacity C is large enough to allow the multi-type level we have chosen.

We can choose the relative frequency of a type in a production cycle proportional to the average demand for that type, or to a rounded value of the average demand. An alternative is to choose the relative frequency equal to the inverse of T_i, the time between set-ups. If we choose for this alternative, we have to be careful and check whether the available capacity is sufficient for each of the types. If the capacity restrictions are not too tight, the best choice seems to be the alternative relative frequency based on the absolute frequency in an uncapacitated situation, because this frequency also considers the average costs. In a situation with a strongly restricted capacity the average costs based on the uncapacitated situation are not very important, whereas the 'traffic intensity' becomes a much more important element in the costs.

The next question considers the orders we have to produce if there is capacity available for a type. Logically, for type i, we will at least produce all orders which have to be delivered within T_i periods. However, if the next production opportunity for type i is more than T_i periods later, it will be profitable to produce all orders which have to be delivered before the next production opportunity, especially if the holding costs for an order are much less than the penalty costs. Of course the number of orders we can produce from a type is bounded by the available capacity. The restricted capacity leads to an extra problem in the situation in which we have to divide the insufficient capacity over several types. In that case we shall use a strategy that Maes and Van Wassenhove (1986) call a South-East strategy, in which we start with the orders that should be delivered by the end of the current period, starting with the type whose next production opportunity is the most far away, then the other types and then we continue with the orders that should be delivered by the end of the next period and so on.

5.2.5. Numerical results

In order to find out whether the extended (x,T)–rule performs well, its performance will be compared with the two-step rule and the semi-fixed cycle rule. In the simulation we will consider three examples, in which we compare the CPU-time as well as the average costs per period. One example has a multi-type level of one, once with tight capacity restrictions and once with a lot of capacity available and a third example in which usually two types of products will be produced in a period and in which the capacity restrictions are rather tight again. In all examples, the orders arrive according to independent Poisson processes.

Example 5.1

In this example we have 3 types of products ($M=3$), a maximum lead time of 4 periods ($N=4$), a capacity $C=12$, set-up time $S=1$, set-up costs $s=50$, penalty costs $p=10$ and holding costs $h=1$. Orders arrive according to independent Poisson processes, with intensity d_{ij} for type i and lead time period j. The matrix D with the elements d_{ij} is given by:

$$D = \begin{bmatrix} 0 & 0.25 & 0.50 & 1.00 \\ 0 & 0.50 & 1.00 & 2.00 \\ 0 & 0.50 & 1.00 & 3.00 \end{bmatrix} \qquad (5.2.17)$$

For the cyclic rules π_i, i=1,2,3, and the corresponding time between set-ups T_i and average costs g_π^i, which we use in the salvage function $L_i(a_i)$ in the first step of the two-step rule, we find the following values:

$$\begin{bmatrix} T_1 & g_\pi^1 \\ T_2 & g_\pi^2 \\ T_3 & g_\pi^3 \end{bmatrix} = \begin{bmatrix} 4 & 17.56 \\ 3 & 21.50 \\ 3 & 22.50 \end{bmatrix} \tag{5.2.18}$$

The average demand per period in this example is 9.75. With one set-up every period we find that about 90 percent (10.75 out of 12) of the available capacity will be used. Therefore we set the multi-type level in the semi-fixed cycle equal to one and the relative frequency proportional to the average demand. We find a cycle with a length of 39 periods, in which the absolute frequency for the types is 7/39,14/39 and 18/39 respectively. In the extended (x,T)–rule we also produce one type in most of the periods, now by choosing $mcl=3$ and ecl=4. We have chosen the matrix of penalty points and the x_i values by a trying some different values based on intuition. Along the lines described in 5.2.2.2., every order has been given penalty points. Of course, the orders with a residual lead time of one period or less, have been given a much larger number of penalty points. For the orders with longer residual lead times, the number of penalty points decreases more or less exponentially. Furthermore, orders for the type with the smallest average demand have been given more penalty points. The x_i values are slightly less than the g_π^i values in (5.2.18). The choices presented in (5.2.19) and (5.2.20) were reached after only two trials. Further changes did not yield better results.

$$Q = \begin{bmatrix} 10 & 4 & 2 & 1 \\ 8 & 3 & 1 & 1 \\ 6 & 3 & 1 & 1 \end{bmatrix} \tag{5.2.19}$$

The pairs (x_i,T_i), i=1,2,3 are chosen as

$$\begin{bmatrix} x_1 & T_1 \\ x_2 & T_2 \\ x_3 & T_3 \end{bmatrix} = \begin{bmatrix} 16 & 4 \\ 20 & 3 \\ 20 & 3 \end{bmatrix} \tag{5.2.20}$$

For the penalty cost functions $P_2(.)$ and $P_3(.)$ we have used different values for mc and α respectively. These values are given between the brackets of the penalty costs function in the Table 5.1.

In this table we see that by making the right choices for α and for mc we can obtain much better results with the penalty cost functions P_2 and P_3 than with the penalty cost function P_1. The increase of the CPU-time is quite small. The average costs of

Strategy	set-up	penalty	holding	total cost	CPU (ms)
P_1	50.75	30.20	4.39	85.34	1973
$P_2(0.5)$	51.55	24.66	5.53	81.74	2086
$P_2(1)$	52.40	18.54	6.49	77.43	2141
$P_2(2)$	56.20	10.16	8.90	75.26	2164
$P_2(3)$	59.85	5.71	11.37	76.93	2201
$P_2(4)$	61.95	4.91	12.24	79.10	2302
$P_3(0.25)$	51.95	21.00	6.09	79.04	2233
$P_3(0.5)$	52.60	18.49	6.63	77.72	2225
$P_3(0.75)$	55.60	11.40	8.21	75.21	2283
$P_3(0.9)$	63.00	4.93	12.50	80.43	2239
ext. (x,T)	51.80	12.65	8.92	73.37	544
cycle	50.00	53.02	9.36	112.38	457

Table 5.1 The performance of the production rules

the semi-fixed cycle rule are very high and this rule does not seem very useful for practical situations. Quite surprising in this example are of course the average costs of the extended (x,T)–rule. We expected that the average costs would be the same as for instance the average costs of P_1, but the average costs are lower than for any other rule. Moreover, the CPU-time is only a little more than the CPU-time for the semi-fixed cycle. The CPU-time for the semi-fixed cycle rule is only based on the generation of the random demand and the administration.

In the P_2 rule with high *mc*-values as well as in the P_3 rule with high α values and the extended (x,T)–rule we are almost not interested in savings on holding costs and therefore we have an advantage of a few orders if we come to a number of periods in which there are severe capacity shortages. We now are curious whether the same elements can be observed in the results of Example 5.2.

Example 5.2

The second example has the same structure as the first one. Now we have 7 product types ($M=7$), a maximum lead time of 5 periods ($N=5$), a capacity $C=30$, set-up time $S=1$ and the same costs: set-up costs $s=50$, penalty costs $p=10$ and holding costs $h=1$. Orders arrive according to independent Poisson processes and the matrix D with the elements d_{ij} is given by:

$$D = \begin{bmatrix} 0 & 0.25 & 0.50 & 0.75 & 0.75 \\ 0 & 0.50 & 0.50 & 0.75 & 0.75 \\ 0 & 0.50 & 0.75 & 0.75 & 1.00 \\ 0 & 0.50 & 0.75 & 1.00 & 1.25 \\ 0 & 1.00 & 1.00 & 1.00 & 1.50 \\ 0 & 1.00 & 1.00 & 1.50 & 1.50 \\ 0 & 1.00 & 1.50 & 1.50 & 2.00 \end{bmatrix} \tag{5.2.21}$$

For the cyclic rules π_i, $i=1,..,7$, and the corresponding time between set-ups T_i and average costs g_π^i, we find the following values:

$$\begin{bmatrix} T_1 & g_\pi^1 \\ T_2 & g_\pi^2 \\ T_3 & g_\pi^3 \\ T_4 & g_\pi^4 \\ T_5 & g_\pi^5 \\ T_6 & g_\pi^6 \\ T_7 & g_\pi^7 \end{bmatrix} = \begin{bmatrix} 4 & 18.31 \\ 4 & 20.25 \\ 3 & 21.00 \\ 3 & 21.50 \\ 3 & 23.83 \\ 3 & 24.33 \\ 3 & 25.33 \end{bmatrix} \tag{5.2.22}$$

The average demand per period in this example is 26.75. With two set-ups every period we have that about 96 percent (28.75 out of 30) of the available capacity will be used. This is rather a high percentage, but with a multi-type level of one the time between two production periods is too long. Therefore we set the multi-type level for the semi-fixed cycle equal to two and the relative frequencies again proportional to the average demands. We find a cycle with a length of 107 periods, in which the absolute frequencies for the types are given by 18/107,20/107,24/107,28/107,36/107,40/107 and 48/107 respectively.

In the extended (x,T)–rule, we now choose $mcl=6$ and $ecl=6$. The matrix Q with the penalty points f_{ij} for type i and residual lead time j is chosen in more or less the some way as in Example 5.1., where orders which cannot yet be produced, and also some of the orders with a small probability for production, are given no penalty point. The best choice is given by:

$$Q = \begin{pmatrix} 10 & 5 & 3 & 1 & 1 \\ 9 & 5 & 3 & 1 & 1 \\ 8 & 4 & 3 & 1 & 0 \\ 8 & 4 & 2 & 1 & 0 \\ 7 & 3 & 2 & 1 & 0 \\ 6 & 3 & 2 & 1 & 0 \\ 6 & 3 & 1 & 0 & 0 \end{pmatrix} \tag{5.2.23}$$

The pairs (x_i, T_i), $i=1,..,7$ are chosen as

$$\begin{pmatrix} x_1 & T_1 \\ x_2 & T_2 \\ x_3 & T_3 \\ x_4 & T_4 \\ x_5 & T_5 \\ x_6 & T_6 \\ x_7 & T_7 \end{pmatrix} = \begin{pmatrix} 22 & 4 \\ 22 & 4 \\ 23 & 4 \\ 24 & 3 \\ 23 & 3 \\ 23 & 3 \\ 24 & 3 \end{pmatrix} \tag{5.2.24}$$

The results of the simulation of this second example are given in Table 5.2.

Strategy	set-up	penalty	holding	total cost	CPU (ms)
P_1	100.75	90.05	15.64	206.44	7123
$P_2(0.5)$	106.05	52.07	21.12	179.24	7646
$P_2(1)$	106.75	48.25	22.45	177.45	7756
$P_2(2)$	112.60	28.27	27.53	168.40	7993
$P_2(3)$	120.00	18.17	33.38	171.55	8074
$P_2(4)$	124.05	14.30	36.86	175.21	8265
$P_3(0.25)$	107.00	52.38	20.91	180.29	8166
$P_3(0.5)$	106.85	46.82	21.96	175.63	8153
$P_3(0.75)$	108.65	40.78	23.60	173.03	8212
$P_3(0.9)$	115.30	24.54	29.58	169.42	8147
$P_3(0.95)$	127.50	12.75	38.22	178.47	8221
ext. (x,T)	110.50	38.74	28.14	177.38	1441
cycle	100.00	139.53	20.19	259.72	1289

Table 5.2 Performance of the production rules

The results in Table 5.2 show a lot of resemblance with the results in Table 5.1: again the average costs of the two-step rule are minimal for good choices of α and *mc* in P_2 and P_3 and again the average costs of the semi-fixed cycle rule are very high. But this time the extended (*x*,*T*)–rule no longer is the production rule with the smallest average costs. We have tried about ten different sets of Q values for this rule and almost every time the average costs per period has been in the interval [177.5,180]. Therefore we expect that we cannot do much better in the extended (*x*,*T*)–rule than the choice we have made. Of course it should be noticed that the CPU-time for this rule is much smaller than the CPU-time for the two-step rule. Without the random demand generator and the administration, the extended (*x*,*T*)–rule may be up to 40 times faster than the two-step rule.

Example 5.3

In the Examples 5.1 and 5.2 the capacity restrictions were very tight. In order to see whether the differences in average costs also appear in the situation in which the capacity is rather loose, we will consider a third example. Therefore we take again Example 5.1, but now with an available capacity of 15 instead of 12. With one set-up every period, about 72 percent (10.75 out of 15) of the available capacity will be used. The cyclic rules π_i are not affected by the change in capacity, therefore (5.2.18) still holds and we can also use the same semi-fixed cycle. For the extended (*x*,*T*)–rule we used the same set of values for the pairs (x_i,T_i) as in (5.2.20). Because it will not be very profitable to produce orders with a longer residual lead time, the penalty points for these orders have been decreased and after some trials we have found the following set of values for the penalty points:

$$Q = \begin{bmatrix} 10 & 3 & 1 & 0 \\ 8 & 2 & 0 & 0 \\ 6 & 2 & 0 & 0 \end{bmatrix} \qquad (5.2.25)$$

We used the values $mcl=6$ and $ecl=6$. In Table 5.3. the results of the simulation of Example 5.3 are given.

In Table 5.3 we see that the differences between the three penalty functions are much smaller than in the previous situations. In this situation, the capacity is usually enough to produce those orders that give a base cost reduction for one of the types, but the remaining capacity is not enough to produce orders for another type. On the aggregated level capacity shortages hardly occur, therefore there is almost no difference between the penalty functions. Only in the situation with $\alpha=0.9$, we

Strategy	set-up	penalty	holding	total cost	CPU (ms)
P_1	46.60	16.49	6.29	69.38	2141
$P_2(0.5)$	46.65	13.32	7.41	67.38	2202
$P_2(1)$	46.70	13.23	7.41	67.34	2209
$P_2(2)$	47.05	12.49	7.60	67.14	2229
$P_2(3)$	48.50	8.15	8.97	65.62	2295
$P_2(4)$	52.85	3.46	11.66	67.97	2505
$P_3(0.25)$	47.10	12.36	7.70	67.16	2283
$P_3(0.5)$	47.45	10.93	7.97	66.35	2301
$P_3(0.75)$	48.80	6.61	9.52	64.93	2271
$P_3(0.9)$	56.90	1.51	14.65	73.06	2379
ext. (x,T)	43.70	11.86	9.81	65.37	542
cycle	50.00	10.22	19.95	80.17	464

Table 5.3 Performance of the production rules

overestimate the future penalty costs enormously and therefore the holding costs and the set-up costs are much higher than with the other choices.

It is not very difficult to draw a conclusion from these examples. The extended (x,T)–rule performs quite well in all examples, using only very little computation time and leading to average costs which are about as good as the average costs for the best choices of α or mc in the two-step rule, especially in situations in which the capacity restrictions are not extremely tight. Finding optimal values for the matrix Q containing the penalty points and for the pairs (x_i,T_i) may seem very difficult, but the production rule is not very sensitive for small changes in these values, allowing us to use almost the same values in both Example 5.1 and Example 5.3. Therefore it will be quite easy to find good values for Q and (x_i,T_i).

5.3. Situation in which extra capacity is available

5.3.1. Introduction

In this section we shall consider the dynamic multi-type, capacitated problem in which extra capacity is available. The extra capacity may be the result of the hiring of extra labour forces for the production unit or it also may be the result of buying products from another firm. In both cases the costs will be higher than the normal production costs. We will assume that the extra costs are linear with the amount of extra capacity that is used. Once again the performance will be measured by the long term average costs per period. This problem is even more complicated than the problem of minimising the average costs in the situation in which we cannot use extra capacity. Therefore we will again concentrate our attention on finding a good heuristic, with low average costs and with small computational efforts.

If there is usually enough capacity available or if the extra costs for extra capacity are very small, then the situation will be much like the situation in Section 4.2. and we can use the heuristics described in that section. In most of the practical situations however, the capacity will be restricted and hiring extra labour forces will be expensive. Therefore we shall consider the same heuristics as in the previous subsection and adapt these heuristics to the possibility of using extra capacity.

Extra capacity can be used to produce some extra orders of a certain product type. Due to the production of some extra orders the next set-up for the type can be delayed and in this way we can save on set-up costs. But the savings due to the production of some extra orders cannot be larger than one extra set-up, at costs s. If the cost for the extra capacity is z per unit of capacity, this implies that it will not be profitable to produce more than $\frac{s}{z}$ extra orders, at least not for the reason of saving on set-up costs. Due to the production of some extra orders we can also save on the penalty costs. This will be profitable for potential late orders, for which penalty costs have to be paid if the orders are not produced by the end of the period. For other orders however, production will only be profitable in the situation in which we are (almost) certain of capacity shortages. If we know that extra capacity will be necessary during the next periods, the extra capacity can be used immediately, but even then it will only be profitable if we can use the capacity for orders that have to be delivered in one of the periods during which capacity shortages occur. Keeping all this in mind, the heuristics from Section 5.2. will be adapted to the possible use of extra capacity.

5.3.2. Extended overtime (x,T)-rule

5.3.2.1. Introduction to the extended overtime (x,T)-rule

In the situation in which there is no possibility for working overtime the extended (x,T)–rule has proven to be a simple and fast heuristic with small average costs per period. Therefore we will adapt this production rule to the possibility of working overtime. This results in the extended overtime (x,T)–rule. There will be only one big difference with the extended (x,T)–rule from Section 5.2.2. and this difference is of course that in some situations we will use more capacity than the normal available amount. The decision about the use of extra capacity will be taken according to a simple rule.

5.3.2.2. Formulation of the extended overtime (x,T)-rule

In this rule we also use penalty points that give an indication for the urgency of an order and to determine the most urgent type. Finding good choices for the penalty points and for the required minimum amount x_i will again be a matter of trial-and-error combined with intuition. Now we will describe the extended overtime (x,T)–rule for a given set of penalty points, with f_{ij}, $i=1,..,M$, $j=1,..,N$, the penalty points for an order of type i with a residual lead time of j periods. In this description we assume that we have a given set of pairs (x_i,T_i), $i=1,..,M$.

The penalty points for a type, written as $K(i)$, are defined as in (5.2.1) by:

$$K(i)=\sum_{j=1}^{N} f_{ij}\, r_{ij} \tag{5.3.1}$$

If $K(i)<x_i$ for all i, then we do not produce at all during the period. Otherwise, we place the types with a sufficient amount of penalty points in a sequence $\{i_1,i_2,..,i_k\}$, with $k\leq M$, such that $K(i_1)\geq K(i_2)\geq..\geq K(i_k)$. Types with an equal number of penalty points are placed according to an increasing average demand. In the uncapacitated situation we would produce all types in this sequence and the amount for type i_l would

be:

$$a_{i_l}=\sum_{j=1}^{T_{i_l}} r_{i_l j} \tag{5.3.2}$$

The first part of the extended overtime (x,T)–rule is now exactly the same as the first part of the extended (x,T)–rule. Following the sequence we allocate capacity to the various types. We do not allocate capacity to the next type if the remaining capacity is less than the minimum capacity level *mcl* and we can possibly produce some extra orders for some of the types if the remaining capacity level is less than the extra capacity level *ecl* and if capacity has been allocated to all product types in the sequence.

The difference with the extended (x,T)–rule happens in the situation in which there is no more capacity available after the first round. Usually this implies that for one of the types, type i say, we have not planned to use the uncapacitated amount a_i+S in the first round but a smaller amount, say C_k. If we have the situation in which only a subset of the orders for period k, with $k<T_i$, are planned to be produced, we will at least produce all orders for type i with a residual lead time of k periods. In all other situations we produce all orders for type i with a residual lead time no longer than T_i, for which the sum of the holding costs and overtime costs is less than the penalty costs for one period:

$$h(\textit{residual lead time in periods})+z\leq p \tag{5.3.3}$$

We will not use extra capacity to produce orders with a residual lead time later than T_i.

5.3.3. The two-step overtime rule

First we shall consider the extension of the the two-step rule described in Section 5.2.3.. Both steps of the heuristic have to be changed if we allow the possibility of working overtime. In the first part we no longer can restrict the number of possible actions for a type by considering only the actions $0,1,..,C-S$. If there are enough orders for a type any action is possible now. Not all actions however, have a chance to be chosen as one of the best actions in the first step. The most obvious candidates for being the best action for a type, are those actions in which we produce all the orders for an integer number of periods and of course, if there are enough orders for the type, also the action $C-S$, in which no extra capacity is used. In order to limit the computation times only these obvious actions are considered. Unlike the marginal penalty costs, the costs for working overtime will be included in the so-called base cost reduction. Therefore the values for the base cost reduction have to be recalculated after every step in the first part of the heuristic.

Except for the case in which all orders will be produced according to the allocations in the first step in the heuristic, we will always come to the second step, even if all normally available capacity has been used. In this second step we consider the production of some extra orders in order to avoid future penalty costs due to capacity shortages. Due to the possible extra capacity we are no longer interested in the sum of the capacity shortages, like we did in the penalty functions P_1, P_2 and P_3, but more in the maximum shortage. Disregarding the variability in the demand it is clear that if we produce an extra amount of orders in this period which is equal to the maximum shortage, there will be no shortages. Therefore we have to use other penalty functions in this situation. Again these functions use the variability of the demand differently:

1) We do not consider any variability, but we just consider the costs for working overtime we would have if the number of orders would equal F_i and if, from the second period on, we would take the maximum action $C_{\max}$. The 'penalty' costs is this model are given by:

$$Q_1(a_a) = z \max_{i=2,\ldots,N} \left[F_i - (i-1) C_{\max} - a_a \right]^+ \qquad (5.3.4)$$

We only have to consider the first N periods, since we assume that $C_{\max} > e_{aN}$.

2) In the second method we will consider the same variability as in the penalty function P_2, by assuming a uniform distribution with the same mean and variance as the original distribution. We have to determine the lower limit and the upper limit for the cumulative number of orders for the first N periods and compare this number with the sum of the available capacity and a_a. Following (5.2.10)-(5.2.12) we write again x_{ai} for the maximum deviation, l_{ai} for the lower limit during period i and u_{ai} for the upper limit. During every period there are three possibilities for the extra overtime costs: if the sum of the available capacity and the aggregated action exceeds the upper limit for the cumulative number of orders, there are no extra costs, if the sum is between the lower and the upper limit, there are some extra costs and if the sum is less than the lower limit, the extra costs for working overtime are complete, that is, like in the

penalty function Q_1:

$$Q_2(a_a) = z \max_{i=2,..,N} \Big((F_i - (i-1)C_{\max} - a_a) 1_{(0,l_{ai}]}((i-1)C_{\max} + a_a) + \\ + \frac{(u_{ai} - (i-1)C_{\max} - a_a)^2 1_{(l_{ai},u_{ai}]}((i-1)C_{\max} + a_a)}{4x_{ai}} \Big) \qquad (5.3.5)$$

We also consider the first N periods, because we cannot produce extra orders for later periods.

3) In this third method we do not consider any variability in the demand, but instead we assume that the penalty function decreases with for every order that is produced. Therefore we also assume 'penalty' costs if the available capacity is large enough. As long as the expected number of orders is larger than the normal available capacity minus set-up times, a multiple of $C_{\max}$, it decreases with an amount z, the costs for 1 unit of extra capacity. If the expected number of orders is k units smaller than the multiple of $C_{\max}$, than the penalty function decreases with an amount $z\,\alpha^k$, where α is a constant between 0 and 1 which can be chosen freely. Again we will consider the period in which the available capacity is the most tight. This results in the following penalty function:

$$Q_3(a_a) = z \max_{i=2,..,N} \Big((F_i - (i-1)C_{\max} - a_a + \frac{1}{1-\alpha}) 1_{(0,F_i]}((i-1)C_{\max} + a_a) + \\ + \frac{\alpha^{((i-1)C_{\max} + a_a - F_i)}}{1-\alpha} 1_{(F_i,\infty)}((i-1)C_{\max} + a_a) \Big) \qquad (5.3.6)$$

The two-step overtime rule is not much different from the two-step rule. In the first part we calculate the base cost for every type and for those actions that consist of producing all orders for a type for an integer number of periods (including 0) or the action in which the normal available completely. This set of possible actions is denoted by A_i, where i is the type of product we consider.

Determine $Co(i,a_i) = q(r_i,a_i) + L_i(a_i) + z(a_{i-C+S}^{+})$ for $i=1,..,M$, $a_i \in A_i$.

We define $Co(i,j) = \infty$ if $a_i \notin A_i$, that is for those actions we expect to be non-optimal or actions which are simply impossible, because there are not enough orders for the type we consider. As in the two-step rule we now determine the base cost reduction $K_0(i,j)$, defined as the difference in the base costs between the use of j units and 0 units of capacity for type i. Using j units of capacity implies that action $a_i = j - S$ is taken. The subscript '0' indicates the number of steps in the algorithm. Let B_i be the maximum use of capacity for type i, that is the sum of S and the largest element of A_i.

Determine $K_0(i,j) = Co(i,j-S) - Co(i,0)$ for $i=1,..,M$, $j=S,..,B_i$.

We define $K_0(i,j) = s$ if $0 < j \le S$ and $K_0(i,0) = 0$ for $i=1,..,M$. Now the capacity 'action' for which the base cost reduction $K_0(i,j)$ is minimal is the best one-step 'action'.

Determine the pair (i,j) for which $K_0(i,j)$ is minimal with $i=1,..,M$.

Assume that the base cost reduction is minimal for the pair (i_0, j_0). If it is optimal to use no capacity, that is if the minimum value is larger than 0, then we continue with the second part of the rule. Otherwise, at least j_0 units of capacity will be used for a set-up and the production of orders of type i_0. Now we set $C_0 = C - j_0$, the remaining capacity, and we adapt the base cost reduction for the type i_0 for the values up to the new maximum B_{i_0} which is defined as the old maximum minus j_0.

Determine $K_1(i_0,j) = K_0(i_0,j+j_0) - K_0(i_0,j_0)$ for $j=0,..,B_{i_0}$.

For the other types the base cost reduction $K_1(i,j)$ can be different from $K_0(i,j)$. If $j \le C_0$ we do not have overtime costs. For the values $j > C_0$ we have extra overtime costs of $\min(zj_0, z(j-C_0))$. If the action $C_0 - S$ does not belong to A_i we also determine the base costs and the base cost reduction for this action.

Determine $K_1(i,j) = K_0(i,j+j_0) + z\min(j_0, j-C_0)^+$ for $j=0,..,B_i$, $i \ne i_0$.

Now we will consider the production for one of the other types.

Determine the pair (i,j) for which $K_1(i,j)$ is minimal with $i=1,..,M$, $j=1,..,B_i$.

As long as we can reduce the base costs we repeat the procedure of determining the best one-step action, adapting the base cost reduction and adapting the remaining capacity. Finally we come to a situation in which the base cost cannot be reduced any more. Then we determine the new values for B_i and we calculate the base costs and the base cost reduction for all values between 0 and B_i, which have not been considered in the first part. The new value for B_i may be different from its previous value, because now we also consider the fact that, regardless of the penalty function that is used, the production of an extra order yields a penalty cost reduction which cannot be larger than the extra costs for one unit of extra capacity. For this reason some of the actions may never be the most profitable ones and therefore the new value for B_i may be smaller than the previous value. In this production rule we always continue with the second part.

In this second part we will consider the possibility of producing some extra orders to avoid capacity problems in the future. Therefore we consider the effect of the aggregated action a_a. Because we have taken already some decisions about the production in the first part, we do not have to consider all possible values for a_a. The lower value for a_a we have to consider is the sum of all actions in the first step of the rule minus the set-up times and is given by $a_{\min}$ in (5.2.15). The upper value we have to consider, $a_{\max}$, is now different from the one that is defined in (5.2.16). Now the upper value is given by the sum of $a_{\min}$ and the maximum amount of extra capacity B_i for all types:

$$a_{\max} = a_{\min} + \sum_{i=1}^{M} B_i \tag{5.3.7}$$

For these values of a_a, we determine the penalty cost function $Q_1(a_a)$, or as an alternative, $Q_2(a_a)$ or $Q_3(a_a)$. Since we are merely interested in the differences in penalty cost values between different values of a_a, we define the marginal penalty costs, written as $Q^*(j)$.

Determine $Q^*(j) = Q_1(a_{\min}) - Q_1(a_{\min}+j)$ *for* $j=0,1,...,a_{\max}-a_{\min}$.

This non-negative function $Q^*(j)$ gives the reduction in penalty costs if we take the aggregate action $j+a_{\min}$ instead of $a_{\min}$. The remaining part of the rule goes along the same lines as before. We determine the pair (i_k, j_k) that yields a minimum value for the extra base costs minus the marginal penalty costs, $K_k(i_k, j_k) - Q^*(j_k)$. If this difference is positive, then we stop. Otherwise, the 'action' (i_k, j_k) will be taken and we will adapt the values for the base cost reduction for all types and we also adapt the marginal penalty costs $Q^*(j)$. Then we continue with the next pair.

Determine the pair (i,j) *for which* $K_{k+1}(i,j)$ *is minimal with* $i=1,...,M$, $j=1,...,B_i$.

Finally, if the extra base costs are larger than the penalty cost reduction, the planned actions can be performed and the algorithm can be repeated the next period.

5.3.4. The semi-fixed cycle rule

The semi-fixed cycle rule does not change very much if there becomes a possibility for working overtime. As well as in the situation with a strict capacity we choose the relative frequency for all types in a production cycle and we will produce the different types according to their sequence in this production cycle. The difference between the rule in the strictly capacitated situation and the rule in this situation is found in the orders that we produce for a type if there is a production opportunity for this type.

Now we always produce at least all known orders that have to be delivered before the next production opportunity. Possibly some extra capacity is necessary for this production. However, if there is still available capacity and if the next production opportunity for the type we produce, type i say, is less than T_i periods later, then it may be profitable to produce some extra orders and thus avoid future capacity problems. This extra production will only be done for those orders that have to be delivered within T_i periods. In the situation in which the available capacity has to be divided over more than one type, this will be done according to the so-called South-East strategy, that has been described in 5.2.4.2.

5.3.5. Numerical results

We will study the performances of the production rules using some examples. Because we also want to study the effect of the possibility of performing work in overtime we shall consider the same situations as in the Examples 5.1 and 5.2 in 5.2.5., once with high costs for working overtime and once with low costs for working overtime.

Example 5.4

In this example the situation from Example 5.1 is considered, with 3 types of products ($M=3$), a maximum lead time of 4 periods ($N=4$), a capacity $C=12$, set-up time $S=1$, set-up costs $s=50$, penalty costs $p=10$ and holding costs $h=1$. The costs for working one extra unit are the same as the penalty costs: $z=10$. Orders arrive according to independent Poisson processes, for which the intensities are given by (5.2.17).

The semi-fixed cycle we use in this Example is the same one as the cycle in Example 5.1. For the extended overtime (x,T)–rule, we use the same penalty points and the same pairs (x_i,T_i) for all types as well as the same values for *mcl* and *ecl*. The results for these choices and for the rules with the penalty costs function Q_1, $Q_2(mc)$ and $Q_3(\alpha)$ are given in Table 5.4..

Strategy	set-up	penalty	holding	overtime	total cost
Q_1	49.70	30.60	3.99	2.27	86.56
$Q_2(4)$	49.60	20.25	5.55	2.76	78.16
$Q_2(6)$	51.15	14.43	6.79	3.31	75.68
$Q_2(8)$	53.85	9.92	7.77	3.52	75.06
$Q_2(10)$	56.00	8.55	8.35	3.83	76.73
$Q_3(0.5)$	48.75	23.64	5.22	3.64	81.25
$Q_3(0.75)$	50.55	17.94	6.38	3.25	78.12
$Q_3(0.8)$	53.30	13.95	6.95	4.19	78.39
ext. (x,T)	51.25	8.50	9.68	3.03	72.46
cycle	50.00	16.81	7.35	9.63	83.79

Table 5.4 The results of the simulation of Example 5.4.

Comparing the results from Table 5.4. with the results from Table 5.1. we find that the effect of the possibility of working overtime only has a very small effect upon the average costs for the different rules, except for the semi-fixed cycle rule. For both Q_1 and Q_3 it would even be better not to use the possibility of working overtime. For Q_2 and for the extended overtime (x,T)–rule the average costs have decreased a little. Obviously the costs of one unit of extra time are too high to make the use of extra time very profitable and we see that in all production rules except in the semi-fixed cycle rule the average use of extra capacity is only about 3 percent of the normal available capacity.

Example 5.5

In this example we will consider the situation from Example 5.4 but now the costs for working one extra unit are much less than the penalty costs: $z=3$. This change has no effect upon the semi-fixed cycle that we use, but for the extended overtime (x,T)–rule we use a different set of penalty points. Due to the small costs for overtime, it gets more important to avoid direct penalty costs than to avoid future penalty costs. Therefore, we decrease the penalty points for orders with a residual lead time of 2 and more periods and we increase the penalty points for the orders with a residual lead time of 1 period or less.

$$Q = \begin{pmatrix} 13 & 3 & 1 & 0 \\ 11 & 3 & 0 & 0 \\ 11 & 2 & 0 & 0 \end{pmatrix} \tag{5.3.8}$$

The pairs (x_i, T_i), $i=1,2,3$ are also chosen differently. The x_i values have been increased to decrease set-up costs.

$$\begin{pmatrix} x_1 & T_1 \\ x_2 & T_2 \\ x_3 & T_3 \end{pmatrix} = \begin{pmatrix} 23 & 4 \\ 26 & 3 \\ 28 & 3 \end{pmatrix} \tag{5.3.9}$$

The best results have been obtained for the values $mcl=4$ and $ecl=4$. This result and the results for the other rules are given in Table 5.5.

Strategy	set-up	penalty	holding	overtime	total cost
Q_1	45.10	13.64	6.08	5.12	69.94
$Q_2(4)$	45.50	12.24	6.47	4.81	69.02
$Q_2(6)$	46.05	10.51	6.98	4.90	68.44
$Q_2(10)$	47.40	6.48	8.79	4.55	67.22
$Q_2(15)$	51.50	3.41	10.26	4.15	69.32
$Q_3(0.75)$	44.90	11.66	7.25	5.31	69.12
$Q_3(0.85)$	45.35	9.26	8.23	5.52	68.36
$Q_3(0.90)$	48.45	5.60	9.45	5.03	68.63
ext. (x,T)	43.65	12.21	8.72	3.26	67.84
cycle	50.00	16.81	7.35	2.89	77.05

Table 5.5 The results of the simulation of Example 5.5.

From Table 5.5. we learn that working overtime indeed can be profitable. The average costs are now much smaller for most of the production rules and the choice of the parameter for Q_2 and Q_3 becomes relatively unimportant. In the two-step rule the use of extra capacity has increased to about 15 percent. Due to our definition of the extended overtime (x,T)–rule and of the semi-fixed cycle rule, the use of extra capacity in these rules is still less than 10 percent. In the following example we will study the effect of flexible capacity on the situation described in Example 5.2.

Example 5.6

This example considers the situation from Example 5.2 with 7 types of products ($M=7$), a maximum lead time of 5 periods ($N=5$), a capacity $C=30$, set-up time $S=1$ and the same costs as in Example 5.4: set-up costs $s=50$, penalty costs $p=10$, holding costs $h=1$ and overtime costs per unit which are equal to the penalty costs: $z=10$. The orders arrive according to independent Poisson processes for the intensities are given by (5.2.21).

The semi-fixed cycle we will use in this example is the same cycle as the one in Example 5.2 with 107 periods. The penalty points for the extended overtime (x,T)–rule and the pairs (x_i,T_i) and the level of *mcl* and *ecl* are also the same as in Example 5.2, not because this would be the best choice, but to illustrate the robustness of this rule. The results for this example are given in Table 5.6.

In Table 5.6. we can recognise the same elements as in Table 5.4. but now even stronger. The results for the penalty functions Q_2 and Q_3 are much worse than in Example 5.2. In Example 5.2. the available capacity was used almost completely by overestimating the future penalty costs. But now this overestimation results in a negative effect, such as unnecessarily high overtime costs. The average costs of the (x,T)–rule are less than in Example 5.2. due to a limited use of extra capacity.

Strategy	set-up	penalty	holding	overtime	total cost
Q_1	100.95	79.47	15.74	2.41	198.57
$Q_2(4)$	104.25	48.11	19.85	5.19	177.40
$Q_2(6)$	105.15	43.87	20.86	7.02	176.90
$Q_2(8)$	104.65	45.70	20.88	7.43	178.66
$Q_3(0.75)$	103.15	49.88	19.94	5.25	178.22
$Q_3(0.85)$	103.15	47.01	20.62	7.12	177.90
$Q_3(0.90)$	104.65	45.70	20.88	7.43	178.66
ext. (x,T)	113.90	25.13	29.87	2.30	171.20
cycle	99.25	50.50	24.25	16.90	190.90

Table 5.6 The results of the simulation of Example 5.6.

Example 5.7

Consider once again the situation of Example 5.6, but with costs for working overtime that are less than the penalty costs: $z=5$. Again the semi-fixed cycle is not affected by this change, but the extended overtime (x,T)–rule is. For this rule we have found the following set of penalty points:

$$Q = \begin{pmatrix} 14 & 5 & 3 & 1 & 0 \\ 14 & 5 & 3 & 1 & 0 \\ 12 & 4 & 3 & 1 & 0 \\ 11 & 4 & 2 & 1 & 0 \\ 10 & 3 & 2 & 1 & 0 \\ 9 & 3 & 2 & 1 & 0 \\ 9 & 3 & 1 & 0 & 0 \end{pmatrix} \tag{5.3.10}$$

The pairs (x_i,T_i), $i=1,..,7$ are chosen as

$$\begin{pmatrix} x_1 & T_1 \\ x_2 & T_2 \\ x_3 & T_3 \\ x_4 & T_4 \\ x_5 & T_5 \\ x_6 & T_6 \\ x_7 & T_7 \end{pmatrix} = \begin{pmatrix} 29 & 4 \\ 30 & 4 \\ 32 & 4 \\ 32 & 3 \\ 32 & 3 \\ 32 & 3 \\ 33 & 3 \end{pmatrix} \tag{5.3.11}$$

and we have found $mcl=6$ and $ecl=8$ as being the best choices for these two levels.

Table 5.7. contains the results of this Example.

Strategy	set-up	penalty	holding	overtime	total cost
Q_1	103.60	56.19	16.82	5.34	181.95
$Q_2(4)$	104.55	39.32	20.51	5.70	170.08
$Q_2(6)$	104.90	34.23	22.09	6.87	168.09
$Q_2(8)$	105.25	35.73	21.87	7.73	170.58
$Q_3(0.75)$	102.65	44.08	20.13	5.97	172.83
$Q_3(0.90)$	105.25	34.39	22.25	7.15	169.04
$Q_3(0.95)$	105.50	33.90	22.25	7.15	168.80
ext. (x,T)	106.50	28.93	28.84	4.41	168.68
cycle	99.25	50.50	24.25	8.45	182.45

Table 5.7 The results of the simulation of Example 5.7.

In Table 5.7. the lowest average costs are about the same as in Example 5.2. From this we can learn that for the more complicated rules such as Q_2 and Q_3, the possibility of working overtime does not have much advantages unless the extra capacity is very cheap. For the simple production rules such as the (x,T)–rule and especially the semi-fixed cycle rule, the use of extra capacity is much more interesting.

Chapter 6

ANALYSIS OF MULTI-TYPE MODELS

6.1. Introduction

In Chapter 5 we have studied multi-type models and we have determined the average costs of various production rules for a few examples by means of simulation. In this chapter we want to consider some simple production rules which can be used to calculate the performance of the production system in certain situations. For example: what can be the expected delivery time for orders of a particular type and priority. The relevance of the production rules is less in the rules themselves than in the information they may give about costs and the elements that contribute to the total costs: penalty-costs, holding costs, set-up costs and costs for working overtime. By using these rules we want to examine which delivery times and which lead times are reasonable. Like in the queuing theory, we do this by accepting orders in the system without specifying a lead time at all on arrival, but by considering the number of periods or the time that they are in the system.

In this chapter we will develop two production rules for two different situations. In both situations we will consider a production situation with one machine on which M different types of products are manufactured. We start with a situation where we have periodic review as in Chapter 4 and Chapter 5. In every period a decision has to be made about the production during that period. Before starting the production of a new type a set-up has to be done. Doing work in overtime is possible. There are not many good production rules for which we can calculate the performance in this situation.

One production rule that allows an almost exact calculation of the performance is the rule in which we produce every product type according to a separate (x,T)–rule, independent of the production of other types. If the capacity restrictions are tight or if the costs for working overtime are not very small, this production rule will not yield very satisfying results. Therefore we will add some of the ideas from Chapter 5 to this rule, in order to make it more sensible. We will call the resulting production rule the multi (x,T)–rule.

The second situation that we will consider is a situation in which we have continuous review: we can change the production plan whenever we like and we can continue the production of a type as long as there are orders for this type. It is not so difficult to think of a production rule for this situation: produce the most urgent type until all the orders for this type have been produced. Of course, a production rule for this situation may also contain other elements, such as a minimum number of orders that is necessary for justifying a set-up, or producing only a maximum number of orders, but the suggested production rule yields a good basis for more elaborated rules. In the production situation we have a restricted capacity and no extra work can be done in overtime. The analysis of the production rule will be executed by using techniques for queuing models. In order to simplify the analysis we will assume an extended service discipline, which means that during the production 'period' of a certain type, we will always produce all the orders for that type, even those arriving during the production. The performance of this rule will be compared with the performance of the fixed production cycle rule.

6.2. The multi (x,T)-rule

As we have already indicated in the introduction of this chapter, this decision rule is based on a separate (x,T)–rule for every type. Some characteristic elements from other rules will be added. For every type the urgency is measured by the number of penalty points. A minimum number of penalty points is necessary before the production of a type is considered. Of course, the order states for the different types will be dependent. We will consider a special form of dependence. Therefore, from the semi-fixed cycle production rule the so-called multi-type level will be added, introduced in 5.2.4. If the number of types with a sufficient number of penalty points is less than the multi-type level, then we produce the orders for all of these types. If the number of types is larger than the multi-type level, we produce a number of types that is equal to this level. By limiting the number of types that can be produced during one period, we can create a special form of dependence. We can also limit the average amount of extra capacity that is needed and of course also the number of set-ups per period.

From the extended overtime (x,T)–rule, introduced in 5.3.2., we will use the element of producing the type(s) with the highest number of penalty points and the element that we will not produce one more type if the remaining capacity is too small. If two types have the same number of penalty points, we will prefer the one with the smallest average demand. One element that remains from the original uncapacitated (x,T)–rule is that we will always produce all known orders of the type that we produce. This element is essential for the simplicity of the analysis.

Resuming the elements described above, the scheduling rule takes the following form: at the beginning of each period we determine the most important type(s), in terms of penalty points, among those types for which the number of penalty points is at least the minimum level. If no such type can be found, there will be no production during that period. Otherwise we will produce all demand of the most important type(s), except the demand that arrives during the production period. The maximum number of types we produce during a period is given by the multi-type level.

The analysis of this rule will be done by using a decomposition approach, which may give a lot of information without too much effort, at least when the multi-type level is very small. The choice of the multi-type level can be done in the same way as we have described in Section 5.2. If the level is too high, we can shorten the length of the periods and thus decrease the level. The analysis of the production rule will be described in the next subsection, where we will use a multi-type level of one and a multi-type level of two.

6.2.1. Analysis of the multi (x,T)-rule

The analysis of this production rule will be done in much the same way as the analysis of the uncapacitated (x,T)–rule. For every type of product we consider a Markov chain in which the state is given by the number of penalty points. The interaction between the different states is found in the transition probabilities. We will consider the situation in which there are orders with different priorities. Every order has a fixed number of penalty points, which will not be changed. None of the orders has a lead time. In order to simplify the analysis, we make the following assumptions:

1) the demand per period per type and priority is integer valued, finite, independent of the demand for other periods or types and stochastically stationary.

2) the penalty points of an order depend on the type and priority only and they are integer valued, preferably with one penalty-point for an order of the lowest priority class, in order to keep the state space as small as possible.

3) the penalty points of a product type is defined as the sum of the penalty points of all orders for that type and a second element. This second element is the product of a type dependent constant C_i and the number of periods that has elapsed since the first arrival of an unfinished order for this type. The number of penalty points of the product type is the states in the Markov chain for that type.

4) the types are ordered according to an increasing average demand, in order to simplify the choice of the type in situations in which two types have the same number of penalty points.

The time that is spent in a state depends on the demand of the type, but also on the probability that the type is produced. According to the production rule we produce a type if the number of penalty points for all other types is smaller (or equal if the index of that type is higher) than the number of penalty points for this type. If the multi-type level L is larger than one, then we produce a type if at least $M-L$ out of the $M-1$ other types have a smaller (or equal) number of penalty points. In our calculation we will use the assumption that the number of penalty points is independent of the number of penalty points for other types. Simulation showed that this assumption does not lead to large errors, especially if the number of types is not too small. We will use this assumption of independence to determine the probability that we produce a type for a given number of penalty points.

In order to calculate the steady state probabilities for a certain state (type, penalty points), we consider one production cycle, just as in the analysis of the uncapacitated (x,T)–rule in 4.2.4.4. In this cycle there is one period in which we produce. Furthermore, we can be in state 0 during some time, in state 1 during some time and so on. The average time that we are in state j during one production cycle will be denoted by $q(i,j)$, where i is the type of product that we consider. Thus $T(i) = \sum_{j \ge 0} q(i,j)$ is the average time between two production periods for type i.

In the description of the calculation of the steady state probabilities, we use the following notation:

b_{ij} is the probability that in an arbitrary period the sum of the penalty points of the orders for product type i arriving in that period equals j.

C_i is a positive integer, indicating the extra penalty points per period if we do not produce type i, while there is demand for this type, independent of the arrival times of the orders.

x_i is a positive integer, indicating the minimum number of penalty points that is needed to consider the production of type i.

$y(i,j)$ is the probability that we do not produce type i if there are j penalty points for this type.

The probability $y(i,j)$ depends on the q values for the other types. Since these values are not known initially, the value for $y(i,j)$ will be determined in an iterative way. Independent of the multi-type level L, we can set for a start:

$$y(i,j)=1 \qquad \text{for } j=0,1,..,x_i-1 \qquad (6.2.1)$$

$$y(i,j)=0 \qquad \text{for } j=x_i, x_i+1,.. \qquad (6.2.2)$$

These choices result in:

$$q(i,0)=\frac{b_{i0}}{1-b_{i0}} \qquad (6.2.3)$$

$$q(i,j)=(1+q(i,0))\,b_{ij}+\sum_{k=1}^{j-C_i} q(i,k)\,y(i,k)\,b_{i,j-k-C_i} \qquad (6.2.4)$$

The first part of (6.2.4) originates from two elements: we can move to a state (i,j) coming from the state $(i,0)$ or from one of the states in which we have produced $(\sum_{j\geq x} q(i,j)(1-y(i,j))=1\ !)$ with probability b_{ij}. In the second part of (6.2.4) we describe that coming from the state (i,k), we can move to the state (i,j) with probability $b_{i,j-k-C_i}$. From the q values the new y values can be determined in the following way.

If L=1 we can determine the new values $y(i,j)$ for $j\geq x_i$, using the assumption that we produce orders from type i if the number of penalty points for the other types is smaller (or equal) or not sufficient for production, by the following formula:

$$y(i,j) = 1 - \prod_{k \neq i} \frac{\sum_{l=0}^{g(k,i,j)} q(k,l)}{T(k)} \tag{6.2.5}$$

where

$$g(k,i,j) = max(j-1, x_k-1) \qquad \text{if } k<i \tag{6.2.6}$$

$$g(k,i,j) = max(j, x_k-1) \qquad \text{if } k>i \tag{6.2.7}$$

If $L=2$ then the new values $y(i,j)$ are given by the probability given in (6.2.5) minus the probability that exactly one of the other types will be produced:

$$y(i,j) = 1 - \left[1 + \sum_{k \neq i} \frac{\sum_{l>g(k,i,j)} q(k,l)}{\sum_{l=0}^{g(k,i,j)} q(k,l)} \right] \prod_{k \neq i} \frac{\sum_{l=0}^{g(k,i,j)} q(k,l)}{T(k)} \tag{6.2.8}$$

with the same definition for $g(k,i,j)$. Like in the extended (x,T)–rule described in Chapter 5, we can extend the rule for a multi-type level of two with another element. If the number of penalty points for the most important type is very large then we produce only one type. In Chapter 5 we have used the maximum capacity level *mcl* for a similar purpose. The idea behind this element is that a large number of penalty points will usually imply that a lot of capacity will be used and that producing a second type would lead to a lot of work in overtime. The maximum number of penalty points for type i for which the production of another type is possible will be denoted by v_i. This new element of the rule has of course also an influence upon the calculation of the $y(i,j)$-values. Now the new values can be determined by:

$$y(i,j) = 1 - \left[1 + \sum_{k \neq i} \frac{\sum_{l=g(k,i,j)+1}^{v_k} q(k,l)}{\sum_{l=0}^{g(k,i,j)} q(k,l)} \right] \prod_{k \neq i} \frac{\sum_{l=0}^{g(k,i,j)} q(k,l)}{T(k)} \tag{6.2.9}$$

Using the new y-values, which we have found in (6.2.5), (6.2.8) or (6.2.9), we determine the new values $q(i,j)$ for $j \geq x_i$ for every type by (6.2.4) and then again the new y-values. This procedure is repeated until the changes in the q- and y-values are negligible. We can limit the state space and thereby the computational efforts by assuming that there is some x_{max} for which: $y(i,j)=0$ for all i and for all $j \geq x_{max}$.

The y-values yield the probability that orders for a particular type will be produced the next period for a given number of penalty points. From these probabilities and the transition probabilities, we can easily calculate the probability that orders for a particular type will be produced in k periods, $k>1$, for a given number of penalty points. From the q-values we can easily calculate the distribution for the number of penalty points at the end of the arrival period of an order of a certain type and priority. Combining this distribution with the probabilities that orders for a particular type will be produced in k periods, yields the distribution for the delivery times for all types and priorities. From the transition probabilities we can also calculate the probability that for a certain type k penalty points correspond with l orders. From these probabilities and the y values we can determine the probability that during one period j orders are produced, which yields the need for extra capacity. The average number of set-ups per period is given by:

$$\sum_{i=1}^{M}(T(i))^{-1} \tag{6.2.10}$$

6.2.2. Numerical results

In this subsection we will consider two examples, with orders of two different priorities, in which we will use the multi (x,T)–rule. In these examples we are interested in the delivery times for the different types and priorities. The performance of the rule will be measured by the set-up costs, the costs for working overtime and by the length of the delivery times. We will assume, arbitrarily, that the revenues will decrease linearly with the length of the average delivery times. If S_{ij} is the average delivery time of an order of type i and priority j and e_{ij} is the average demand per period for these orders and the revenue for an order of type j is R_j, the average revenues per period are given by

$$R=\sum_{i=1}^{M}\sum_{j=1}^{2} R_j\, e_{ij}\,(5-S_{ij}) \tag{6.2.11}$$

In both examples we will assume a geometrical distribution for the demand. To simplify the analysis, we will set a maximum for the number of orders per period for a certain type and priority, but neither this maximum nor the distribution will be an essential element in the analysis. We will assume that a set-up takes one unit of time. In the examples we will measure the performance of the rule with a given set of penalty points in three situations with different costs:

a) set-up costs $s=100$, costs for working overtime $z=20$ per unit and the reward per order $R_1=10$ and $R_2=30$.

b) set-up costs $s=100$, costs for working overtime $z=100$ per unit and the reward per order $R_1=10$ and $R_2=30$.

c) set-up costs $s=100$, costs for working overtime $z=20$ per unit and the reward per order $R_1=20$ and $R_2=30$.

Example 6.1.

In this first example we shall consider a situation with four types of products ($M=4$), with an average demand per period that is given by

$$\begin{bmatrix} e_{11} & e_{12} \\ e_{21} & e_{22} \\ e_{31} & e_{32} \\ e_{41} & e_{42} \end{bmatrix} = \begin{bmatrix} 0.900 & 0.100 \\ 0.750 & 0.250 \\ 2.250 & 0.250 \\ 1.875 & 0.625 \end{bmatrix}$$

For this example, we will only consider an (x,T)–rule with a multi-type level $L=1$ and with no extra penalty points per period, $C_i=0, i=1,2,3,4$. The choices for the penalty points for an order of type i and priority j are denoted by f_{ij}. We will consider three different choices for the matrices XF, which contains the values for the minimum level and the penalty points for the different types.

$$XF_1 = \begin{bmatrix} x_1 & f_{11} & f_{12} \\ x_2 & f_{21} & f_{22} \\ x_3 & f_{31} & f_{32} \\ x_4 & f_{41} & f_{42} \end{bmatrix} = \begin{bmatrix} 4 & 2 & 9 \\ 4 & 2 & 8 \\ 4 & 1 & 7 \\ 6 & 1 & 5 \end{bmatrix}$$

and also

$$XF_2 = \begin{bmatrix} 5 & 1 & 5 \\ 5 & 1 & 5 \\ 5 & 1 & 5 \\ 5 & 1 & 5 \end{bmatrix}, \quad XF_3 = \begin{bmatrix} 1 & 1 & 5 \\ 1 & 1 & 4 \\ 1 & 1 & 3 \\ 1 & 1 & 2 \end{bmatrix}$$

For the calculation of the average amount of work in overtime we will assume that the normal available capacity will be 8 units of time, implying that the extra capacity is only needed for the variations in the capacity requirements. The analysis of this example for the various sets of penalty points provided us with much useful information, such as the probability that we produce 6 units of products in one period, the percentage of the orders of type 1 and priority 2 with a delivery time of 3 periods, the average time between two production periods of type 4 and so on. This information will be summarised in Table 6.1.

XF-values	XF_1	XF_2	XF_3
S_{11}	2.81	3.85	3.22
S_{21}	2.66	3.44	3.01
S_{31}	2.42	2.16	1.93
S_{41}	2.23	1.95	1.95
S_{12}	1.68	2.23	1.82
S_{22}	1.66	2.02	1.94
S_{32}	1.61	1.58	1.70
S_{42}	1.62	1.39	1.80
set-up costs	98.82	96.50	99.99
overtime	1.60	1.43	1.27
profit a	140.30	141.96	148.98
profit b	12.57	27.28	47.03
profit c	287.55	285.10	306.19

Table 6.1 The (x,T)–rule *in Example 6.1.*

In Table 6.1. we see that the average delivery times for the different types and priorities can be very different. By making other choices for *XF* we can obtain various ratios for the average delivery times. Considering the profit, XF_3 is the best choice in all three profit situations. By using these penalty points, the amount of work in overtime is limited, a set-up is done every period and the average delivery times are about the same for all types and priorities, except for the orders of priority one of the types 1 and 2.

Example 6.2.

In this second example we consider a situation with six types of products ($M=6$). Again, we will consider an (x,T)–rule with a multi-type level $L=1$ and with no extra penalty points per period, $C_i=0, i=1,2,3,4.$, but we will also consider some choices with a multi-type level of two, in combination with extra penalty points and a maximum level, v_i for the production of more than one type in a period. The average demand per period in this example is given by

$$\begin{bmatrix} e_{11} & e_{12} \\ e_{21} & e_{22} \\ e_{31} & e_{32} \\ e_{41} & e_{42} \\ e_{51} & e_{52} \\ e_{61} & e_{62} \end{bmatrix} = \begin{bmatrix} 0.900 & 0.100 \\ 0.750 & 0.250 \\ 2.250 & 0.250 \\ 1.875 & 0.625 \\ 4.500 & 0.500 \\ 3.750 & 1.250 \end{bmatrix}$$

We will consider one choice for the matrix XF for the multi-type level $L=1$ and three choices for the matrices $XFCV$ for $L=2$. The matrices $XFCV$ contain the values for the minimum level x, the penalty points for the different types and priorities q, the extra penalty points per period C and the maximum level v that allows another production. We will consider the following choices:

$$XF = \begin{bmatrix} 1 & 1 & 7 \\ 1 & 1 & 6 \\ 1 & 1 & 5 \\ 1 & 1 & 4 \\ 1 & 1 & 3 \\ 1 & 1 & 2 \end{bmatrix}$$

and

$$XFCV_1 = \begin{bmatrix} x_1 & f_{11} & f_{12} & C_1 & v_1 \\ x_2 & f_{21} & f_{22} & C_2 & v_2 \\ x_3 & f_{31} & f_{32} & C_3 & v_3 \\ x_4 & f_{41} & f_{42} & C_4 & v_4 \\ x_5 & f_{51} & f_{51} & C_5 & v_5 \\ x_6 & f_{61} & f_{62} & C_6 & v_6 \end{bmatrix} = \begin{bmatrix} 2 & 2 & 7 & 3 & 45 \\ 2 & 2 & 6 & 2 & 44 \\ 1 & 1 & 5 & 2 & 30 \\ 1 & 1 & 4 & 1 & 30 \\ 1 & 1 & 4 & 1 & 28 \\ 1 & 1 & 3 & 1 & 28 \end{bmatrix}$$

and also

$$XFCV_2 = \begin{bmatrix} 1 & 1 & 5 & 1 & 30 \\ 1 & 1 & 5 & 1 & 30 \\ 1 & 1 & 5 & 1 & 30 \\ 1 & 1 & 5 & 1 & 30 \\ 1 & 1 & 5 & 1 & 30 \\ 1 & 1 & 5 & 1 & 30 \end{bmatrix} , \quad XFCV_3 = \begin{bmatrix} 2 & 2 & 9 & 3 & 45 \\ 2 & 2 & 8 & 3 & 44 \\ 4 & 1 & 8 & 2 & 30 \\ 4 & 1 & 6 & 1 & 30 \\ 6 & 1 & 6 & 1 & 28 \\ 6 & 1 & 4 & 1 & 28 \end{bmatrix}$$

For the calculation of the average amount of work in overtime we will assume that the normal available capacity will be 19 units of time. This is the average capacity requirement if we have two set-ups per period. The information of the analysis with the different values for *XF* and *XFCV* will be summarised in Table 6.2. First we will give the average delivery times for the orders from the different types and the average delivery time for the orders with priority one and for the orders with priority two. Then we will give the set-up costs, the average amount of extra capacity that is used and finally the profit in the three different situations.

$XFCV$-values	XF	$XFCV_1$	$XFCV_2$	$XFCV_3$
$\overline{S}_{1.}$	6.63	2.07	3.21	2.48
$\overline{S}_{2.}$	5.24	2.11	2.84	2.31
$\overline{S}_{3.}$	3.64	2.05	2.17	2.36
$\overline{S}_{4.}$	3.15	2.03	1.92	2.24
$\overline{S}_{5.}$	2.48	1.68	1.63	1.91
$\overline{S}_{6.}$	2.42	1.58	1.44	1.77
$\overline{S}_{.1}$	3.24	1.86	1.94	2.12
$\overline{S}_{.2}$	2.64	1.55	1.48	1.66
set-up costs	100.00	192.72	188.62	172.80
overtime	2.16	2.51	2.36	2.46
profit a	313.82	505.46	507.48	479.88
profit b	141.04	304.60	318.57	283.07
profit c	560.19	946.13	936.31	883.31

Table 6.2 The (x,T)–rule *in Example 6.2.*

In Table 6.2. we can observe that a multi-type level of one leads to average delivery times that are much longer then the delivery times with a multi-type level of two. Only if the costs for a set-up would be very high a multi-type level of one can be profitable. The three choices for *XFCV* show some different configurations for the average delivery times. In order to obtain the largest profit it appears to be necessary that the values for the minimum levels x_i are very small. Obviously, waiting for more orders only makes sense if the set-up costs are very high.

6.2.3. Other possibilities and conclusions

If we are not satisfied with the outcome of the delivery times or the profit, we can change:

x_i increasing x_i leads to a less frequent production of type i and less set-ups, decreasing x_i to a more frequent production and usually shorter delivery times.

C_i increasing C_i leads to a more frequent production of type i. Increasing C_i for all types at the same time leads to a more cyclic production pattern.

v_i increasing v_i (if $L = 2$) makes it more often possible to produce more than one type of product in one period. This leads to shorter delivery times, but also to an increasing amount of work in overtime.

f_{ij} by changing the penalty points of the different priorities and different types, the delivery times can get nearer to the desired value.

Other possible situations to analyse are for instance the change of the length of the period, the possibility to refuse orders if the number of penalty points is higher than some maximum, producing different types in one period or on different machines and using and estimating the (negative) correlation between the penalty points of different types. It should be mentioned that this method of approximative analysis is not always faster than simulation, especially not if $L = 2$ and if the number of possible states is large.

6.3. Continuous review model

6.3.1. Introduction

Until now, we have assumed that the service time of an order and the set-up time can be expressed in an integer number of units of capacity. We also have assumed that a set-up has to be done if a type is manufactured during a period and that the decisions about the production and the delivery times are taken periodically, for instance every day or every C time units. In this section we will consider a situation with continuous review: every time that the order state changes, either by the production of an order or by the arrival of a new order, we can take a decision about the production or about the delivery times.

In some production processes the lifetimes of some parts of the machine are given. After such a time interval these parts have to be replaced, cleaned or repaired. In other production processes there is a working shift of for instance eight hours and after this time the machine is stopped. The next working day the machine has to be started and this may take some time. In these situations, periodic review makes sense. It can also be sensible because of organisational reasons. However, in other cases it may be profitable to have a continuous review planning system. In this way, we can use other service disciplines than normal gating service. Now we can consider extended service, in which continuation of the production of a type may take place until all orders have been produced, or even extra extended service, in which we can delay the set-up for the next type and wait for further orders of the type we have produced.

In order to analyse a situation with continuous review and extended service, we assume that the orders for the different types arrive according to independent Poisson processes. Furthermore we assume, for simplicity, that the service time of an order is exponentially distributed. We consider the situation with one machine on which M types of products can be produced, each with a service rate μ_i and an arrival rate λ_i, (i=1,..,M). The set-up time is also exponentially distributed with mean s^{-1}. We also assume that all orders have the same priority and that the available capacity and speed is large enough, that is:

$$\sum_{i=1}^{M} \frac{\lambda_i}{\mu_i} < 1 \tag{6.3.1}$$

In this situation we want to study two service disciplines in combination with a simple production decision rule. Similar to the analysis of the (x,T)–rule, we will not consider lead times, but instead delivery times that are realised with the production decision rule. We will measure the performance by considering the average delivery times for the different types of products and we will assume that the revenues will depend linearly on these average delivery times. In the performance we will also consider the set-up costs. We do not have holding costs, penalty costs or costs for working overtime. First we describe the production decision rule in combination with the extended service and consider the solution: an approximative method to determine the average delivery time for the various types of products for different values in the production decision rule.

6.3.2. Production model

Also in the continuous review situation a natural element of the production scheduling is the clustering of orders for the same type. Every time a certain type is produced, we will produce all demand for that type in order to avoid set-ups. Another element that is quite obvious is that we will schedule the most important or most urgent type first. The most important type is the one for which the number of orders, or the number of penalty points, is the highest among all types. We can use these penalty points as an instrument for controlling the delivery times, but the use of penalty points leads to a larger and more complex state space. Therefore we will consider another instrument for the control of the delivery times. This production rule shows much resemblance with the (x,T)–rule: we will only start the production of a type, type i say, if the number of orders for that type equals at least a minimum level, x_i. By this rule we can both limit the number of set-ups and favour the most profitable types.

Summing up, the scheduling rule takes the following form: each time when the production of a type is finished, we determine the most important type among those types for which the demand is at least the minimum level x_i. If there is such a type then we produce all demand of the most important type, including the demand that arrives during the production. Otherwise, if no such type can be found, the reaction depends upon the service discipline. In the situation with normal extended service, we wait until the demand for one of the types is at least the minimum level x_i. If the first type for which this minimum level is reached is the same as the type that was produced previously, a new set-up is necessary. In the situation with extra extended service we will also wait for further demand, but now this may lead to a continued production of the type that was produced last without doing a set-up. Usually a firm cannot choose freely between normal extended service and extra extended service, for instance if the products are manufactured in an oven. Therefore we will consider both service disciplines.

The purpose of this production model is not to arrive at optimal production rules, although it seems reasonable to assume that an optimal scheduling system will show much resemblance with this model. Our production model however, will give information about the average delivery times that are possible for the different types and information about the number of set-ups and about the required capacity. In this subsection we limit ourselves to the situation in which the importance of a type is solely measured by the number of orders, whereas in some optimal scheduling model the arrival date of an order, the required capacity and the urgency of an order will also be important elements in the planning. Because we will base the decisions only upon the number of orders for the different types, we simply have to solve one problem in our production model, namely to determine the optimal value of the vector $X=\{x_1,..,x_M\}$, which contains the minimum levels for the production of the different types.

Determining the optimal value of X exactly by means of analysis will be impossible in complex situations. Simulation studies on the other hand may be very time consuming. Therefore we will describe a decomposition approach, which may give much of the required information without too much effort. We start with the extra extended service discipline. The results for the normal extended service discipline are then easily found by substitution in the formulae that will be obtained for the extra extended service discipline.

6.3.3. Decomposition model

In the decomposition model, we will consider each type separately, using the following approximations of the scheduling model:

1) if for type i the minimum number of orders, x_i, is reached, it will take an exponentially distributed time, with average $(b_i\mu_i)^{-1}$, before the production of the type starts. This time includes a set-up and the waiting for the production of other types that will be produced before type i;

2) if the production of type i is finished, the probability that for no other type the demand is sufficient, is c_i;

3) if the production of type i is finished and the demand for the other types is insufficient, it will take an exponentially distributed time, with average $(d_i\mu_i)^{-1}$, before the demand of one of the other types reaches its minimum.

Using these approximations, we can model the demand for each type separately as a continuous-time Markov chain. Let us consider one type, with arrival rate λ, service rate μ and a production minimum x. In the Markov chain two elements are playing a role: the number of orders for the type and the state of the machine. The machine can be set for the production of the type or not set for the production. The states will be denoted by k or k^*, where k denotes the number of orders for the type and * indicates that the machine is ready to produce orders for the type. The steady-state probabilities for the states will be denoted by p_k or p_k^* respectively. We now have to solve the following set of equations:

$$\lambda p_0=\mu(dp_0^*+(1-c)p_1^*) \tag{6.3.2}$$

$$\lambda p_k=\lambda p_{k-1} \qquad k=1,2,..,x-1 \tag{6.3.3}$$

$$(\lambda+b\mu)p_k=\lambda p_{k-1} \qquad k=x,x+1,... \tag{6.3.4}$$

$$(\lambda+d\mu)p_0^*=c\mu p_1^* \tag{6.3.5}$$

$$(\lambda+\mu)p_k^*=\lambda p_{k-1}^*+\mu p_{k+1}^* \qquad k=1,2,..,x-1 \tag{6.3.6}$$

$$(\lambda+\mu)p_k^*=\lambda p_{k-1}^*+\mu p_{k+1}^*+b\mu p_k \qquad k=x,x+1,... \tag{6.3.7}$$

The states and the traffic intensities for this set are given in the following figure:

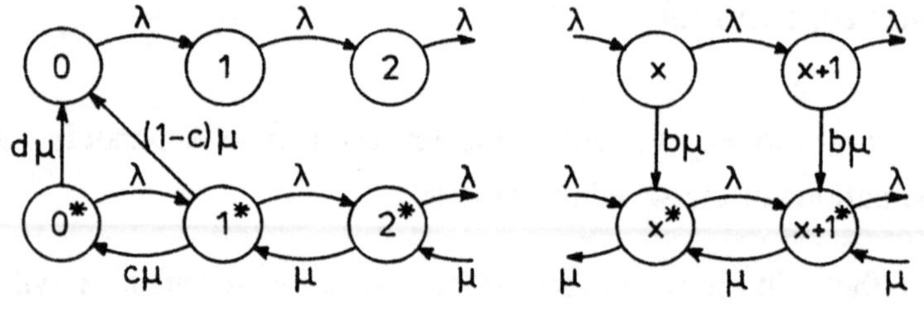

Figure 6.1. The possible states for one type in the production model.

Solving this system yields the following solution:

$$p_0=\frac{b(1-\rho)(d+\rho(1-c))}{(xb+\rho)(d+\rho(1-c))+bc\rho} \tag{6.3.8}$$

and for the average number of orders of the type in the queue:

$$L=\frac{p_0}{1-\rho}\left[\frac{x(x+2\rho-1)}{2}+\frac{x\rho^2}{1-\rho}+\frac{x\rho}{b}+\frac{\rho^2}{b(1-\rho)}+\frac{\rho^2}{b^2}+\right.$$
$$\left.+\frac{\rho^2 c}{(1-\rho)(d+\rho(1-c))}\right] \tag{6.3.9}$$

where $\rho=\frac{\lambda}{\mu}$. Since we have Poisson-arrivals the average delivery time is given by

$$S=\frac{L}{\lambda} \tag{6.3.10}$$

and the set-up rate by

$$u=\lambda p_0 \tag{6.3.11}$$

Of course the choice of b, c and d is very important for the accuracy of this model. In the examples that we will consider, the demand rates for the various types are not very different. We have tried several formulae for b and compared the results of the approximative analysis with simulation results for a large number of examples. According to these examples, the best choice for b_i, the b value for type i, is given by:

$$b_i=[\mu_i(\sum_{j\neq i}w_j+s^{-1})]^{-1} \tag{6.3.12}$$

where w_j is a measure for the waiting time due to the production of orders for type j, were we count all orders for type j, except for the orders in the states $0,1,2,x-1$:

$$w_j=\frac{(L_j-0.5x_j(x_j-1)p_{0j})}{\mu_j-\lambda_j} \tag{6.3.13}$$

Due to the assumption of independence, an obvious choice for c_i is found by the product of the probabilities for each of the types to be in one of the states $0,1,..,x-1$:

$$c_i=\prod_{j\neq i}x_j p_{0j} \tag{6.3.14}$$

The choice for d_i is based on the sum of the transitions rates, to the state x_j for type j. We have multiplied the transition rate with $1/x_j$, which is the probability to be in the state x_j-1, under the assumption of an equal probability for each of the states $0,1,..,x_j-1$, all for type j.

$$d_i=\frac{\sum_{j\neq i}(\frac{\lambda_j}{x_j})}{\mu_i} \tag{6.3.15}$$

For a given set of $\{x_1,..,x_M\}$ the values of b_i, c_i and d_i will be determined by means of iteration, because their value depends on the value of p_{0i} and L_i.

6.3.4. Results for the normal extended service

If we have a normal extended service discipline, then we can simplify the formulae obtained in the previous subsection. In this service discipline we cannot wait for further orders of a type if we have produced all the orders for this type. This implies that we cannot go from the state 1^* to the state 0^*; if the last order is finished a set-up will be necessary before we can continue the production. Usually we will produce one or more other types of products before we start a set-up for the type that is just finished. The effect of this on the formulae is very simple: by substituting $c_i = 0$ for every type i, we can find the formulae for the normal extended service discipline. By choosing this value for c_i, we will also find that the elements containing d_i disappear completely from the formulae. This is of course not a surprise, because the state 0^* simply disappears.

The probability that there are no orders for a type with a given ρ and x is now given by:

$$p_0 = \frac{b(1-\rho)}{xb+\rho} \tag{6.3.16}$$

and for the average number of orders of the type in the queue we find:

$$L = \frac{b}{xb+\rho}\left[\frac{x(x+2\rho-1)}{2} + \frac{x\rho^2}{1-\rho} + \frac{x\rho}{b} + \frac{\rho^2}{b(1-\rho)} + \frac{\rho^2}{b^2}\right] \tag{6.3.17}$$

From this probability and from the average number of orders in the queue, we can calculate the average delivery time and the set-up rate. Again the choice of b_i is important, but we have found that we can use the same approximation as in (6.3.12) and determine b_i by means of iteration.

6.3.5. Numerical results and comparison with fixed cycle

In the examples we will consider in this section, we want to maximise the profit. The elements we consider in this arbitrary profit function P, are the average delivery times S_i for the various types and the set-up rates, $u_i = \lambda_i p_{i0}$, for the various types:

$$P = \sum_{i=1}^{M} (5\lambda_i(2-S_i)-20u_i)$$

In order to maximise this profit, we will try different sets of $\{x_1,..,x_M\}$, for instance starting with increasing the x for the types with the smallest λ or decreasing the x for the types with the largest λ.

We will compare this production model with the extended gating cyclic service model that has been described in Subsection 3.2.1. In a production cycle we have a fixed time T_i available for the production of type i including one set-up. Sometimes this time may not be used entirely, but at other times the time will not be enough to produce all orders, leading to orders that have to wait until the next cycle. By means of iteration, we can determine the optimal values for $T_i, i=1,..,M$ and the corresponding values for the average delivery time of the orders and the profit.

Now we will compare the results of the decomposition method, with normal extended service and with extra extended service, with the fixed cycle. For two different examples we will determine the set of values X_{opt} with the highest profit. We will compare the profit with the maximum profit for the fixed cycle production rule.

Example 6.3.

In the first example we will consider three different types of products, for which the arrival rates and service rates are given by:

$$\begin{bmatrix} \lambda_1 & \mu_1 \\ \lambda_2 & \mu_2 \\ \lambda_3 & \mu_3 \end{bmatrix} = \begin{bmatrix} 5 & 18 \\ 4 & 15 \\ 3 & 12 \end{bmatrix}$$

The expected time for a set-up is 1/18. In Table 6.3. we will give the choices for X_{opt} which give the optimal set according to the decomposition method, as well as the optimal set of available production times in the cyclic model. For these optimal values we will give the average delivery times and the profit P.

	X_{opt} or T_{opt}	S_1	S_2	S_3	profit
extra ext.	{3,3,3}	0.67	0.73	0.83	42.29
normal ext.	{3,3,3}	0.72	0.78	0.87	39.64
fixed cycle	{0.99,0.95,0.90}	1.22	1.39	1.65	15.88

Table 6.3 The results of Example 6.3.

Example 6.4.

In the second example we will consider four different types of products, for which the arrival rates and service rates are given by:

$$\begin{pmatrix} \lambda_1 & \mu_1 \\ \lambda_2 & \mu_2 \\ \lambda_3 & \mu_3 \\ \lambda_4 & \mu_4 \end{pmatrix} = \begin{pmatrix} 5 & 18 \\ 4 & 15 \\ 3 & 12 \\ 2 & 18 \end{pmatrix}$$

The expected time for a set-up is again 1/18. In Table 6.4. we will give the choices for X_{opt}, which give the optimal set according to the decomposition method and also T_{opt}, the optimal set of available production times in the cyclic model. For these optimal values we will give the average delivery times and the profit P.

	X_{opt} or T_{opt}	S_1	S_2	S_3	S_4	profit
extra ext.	{2,2,1,2}	1.16	1.20	1.20	1.49	16.68
normal ext.	{2,1,1,2}	1.15	1.15	1.19	1.47	16.20
fixed cycle	{1.88,1.81,1.71,0.82}	3.38	3.87	4.64	5.71	-161.45

Table 6.4 The results of Example 6.4.

From these tables we may conclude that the difference between the extra extended service and the normal extended service is very small, but that the difference between the fixed production cycle and the other production rule is very large, especially if the system is heavy loaded, as in Example 6.4. The length of the fixed cycle has to be very long to avoid that too much time is lost on set-ups. In a heavily loaded situation there will nearly always be a sufficient number of orders for one of the other types if the production of a type is finished and therefore the difference between the two ways of extended service disappears almost completely.

The purpose of analysing situations in this way is to get an impression of the average delivery times that are possible for the various types of products and also an impression of the number of set-ups per time unit. The average delivery times may be important for the profit like in the examples above and they can influence the demand and the profit. From Examples 6.4 we can learn that it would be better not to have the orders for type 4, because they only lead to a decreasing profit. Starting from the analysis of the simplified situation we can add other elements, such as orders with different priority, a realistic distribution for the arrival of orders and for the service times and we may consider decision rules for lead times. Some of these extensions can be treated with an analysis similar to the one that we have described or by using a so-called Mean Value technique. For other extensions the performance can only be measured by a simulation study.

Chapter 7

THE (x,T)-RULE IN COMPLEX SITUATIONS

7.1. Description of the problem

In Chapter 4 we have introduced the (x,T)–rule for an uncapacitated situation. In Chapter 5 we have extended the rule for a multi-type capacitated situation with fixed lead times. In this chapter we want to study the necessary extensions for the use of the (x,T)–rule in more complex situations with firm-initiated lead times. The situations in this chapter are combinations of the more simple situations that we have considered in the previous sections: several product types with different demand rates, set-ups between types, orders with different priorities, backlogging, capacity constraints and overtime possibilities.

The different priorities of the orders imply that the reaction upon the proposed lead times will be different and that the penalty costs and the revenues will be different. The decisions about the lead times and about the production are made periodically, but different from the periodic review models in Chapter 4 and Chapter 5, the production of a type can be continued in the next period without an extra set-up at the beginning of a period. This assumption allows us to treat non-stop production processes in the same way as production processes with only one or two working shifts per day. It will be required that the average amount of capacity that is used for working overtime will be less than a few percent of the fixed capacity that is available and it will also be required that the number of late orders is limited, for instance no more than a few percent for the most urgent orders. Due to this assumption it is not unreasonable to assume that the demand is not influenced by the lateness of the previous orders, so that we can consider a stationary demand process.

The lead times that are proposed by the firm will be based on a preliminary production plan. The lead times for the orders of different priorities will be bounded by a maximum number of periods, which depends on the priority, thus offering the clients some certainty. Of course, the lead times have to be met quite accurately.

First we will consider the use of an extended (x,T)–rule in the situation containing the elements described above for one machine and assuming that all lead times are accepted by the clients. This situation will be extended by assuming probabilities that the lead times for the orders are accepted, just as in Section 4.4. These probabilities will depend upon the priority and the lead time that is offered for the order. The second extension which we will consider is the extension to different machines. Different types of products can be produced on different machines, some types on only one machine and other types on two or more machines. This complicates the production planning and the rules for the lead times, but we will notice that applying an extended (x,T)–rule is still quite simple and leads to a good performance.

7.2. The elements of the extended (x,T)-rule

The extended (x,T)–rule for this situation contains a lot of well-known elements from the various (x,T)-rules described in the previous chapters. Again, the basis is a separate (x,T)–rule for every type which says that we can produce the orders which have to be delivered within T periods if there is a sufficient number of penalty points for the type. Due to the firm-initiated lead times it will often happen that the due date is the same for almost all the orders for a specific type. The first orders for the type obtained for instance a lead time of 4 periods, the orders arrived in the next period a lead time of 3 periods, then 2 periods and so on. This reduces the importance of the choice of T quite a lot, because usually most of the orders that are produced are orders that have to be delivered by the end of the period and the number of orders that will be produced will depend more on the capacity limitations than on the choice of T. The choice of the penalty points is important for the differences in delivery times for the various types and priorities and for the set-ups. This choice will be described more closely in the next subsection. If there are more types for which the number of penalty points is sufficient for production, we will choose the type(s) with the highest number of penalty points. If two types have the same number of penalty points we will prefer the type with the smallest average demand.

Another element of the (x,T)-rule for this situation is the maximum number of types that will be produced on one machine during one period, the so-called multi-type level L. This level has been introduced to avoid that the number of set-ups will be too large. In this chapter it will merely be used in the preliminary production plan to avoid capacity problems in future periods. An element that is related with the multi-type level is the minimum capacity level, introduced as *mcl* in Chapter 5. If we have allocated capacity to one or more types and the remaining capacity is less than *mcl*, we do not allocate this capacity to another type. In Chapter 5 this element has been introduced because in the situation in that chapter we assumed that it was better to use the capacity for production than for a set-up. In this situation we can continue the production in the next period. Therefore the set-up is not wasted. However, in the next period we will have a lot of new information and therefore it may be better to wait with the production until the beginning of the next period instead of starting it at the end of period. In a practical situation a lot of the new information may be present at the end of the period and then a set-up can be started.

7.2.1. The penalty points

The penalty points are intended to give a measure for the urgency of the production of a type or the urgency of the production of a certain order. In a simple situation this urgency can be measured by estimating the future costs, but in a more complex situation this will be impossible and we can consider the penalty points to be decision variables. We can choose the penalty points for the orders in different ways: in Chapter 4 every order that should be delivered by the end of the first period has 1 penalty point, in Chapter 5 the number of penalty points can take several values, depending on the time until the due date and in Chapter 6 the number of penalty points for an order does not change at all, although this choice has only been made to simplify the analysis. In our opinion a very reasonable way to choose the penalty points for an order in a practical situation is the following one. Like in the (x,T)-rules described in Chapter 5, the penalty points of an order will depend on the residual lead time. The number of penalty points for an order is a constant, q_1, until one period before the due date d. This one period has proven to be sufficient in order to avoid large penalty costs. Furthermore, it yields little holding costs. From the period before the due date on, the penalty point function increases rapidly to avoid that the order will be much too late. Therefore, from the due date on this constant is multiplied every period with a constant q_2, with $q_2 > 1$. This results in the following function for the penalty points of an order if we consider the production in period t:

$$pp\,(order) = q_1 \cdot q_2^{(t-d+1)^+} \tag{7.2.1}$$

An advantage of using this formula is that we can describe the penalty points for an order of a certain type and priority by two parameters q_1 and q_2 instead of a larger vector containing a different value for every value of the residual lead time $(d-t-1)$. For every type, two extra elements will be included in the penalty points. These elements will be described in the next paragraph. The choice of the values for q_1 and q_2 has to be made by trying several values in a simulation study, although some elements are already clear beforehand. Orders for types with a small average demand can be given more penalty points in order to keep their delivery times acceptable and orders with a higher priority can also be given more penalty points to ensure that their delivery time is not too long. If the situation is not too complicated we can first do an approximation analysis as in Subsection 6.2. and use the values from this analysis as a start. We will illustrate this in Example 7.1.

The penalty points for a type will be the sum of the penalty points of all the orders for the type and an additional constant for every period after the last production period in which there has been orders for this type. This is similar to the production model in Subsection 6.2., where we have introduced the extra constant for type i as C_i. In the model in this chapter the extra points are lost as soon as the production of the type has been started, even if not all the orders for the type have been produced. In order to promote a further production of the type for which not all of the orders have been produced in the previous period one more element is added to the penalty points for a type. This element is the sum of the penalty points of all the orders for the type which have to be delivered by the end of the first period multiplied by a constant which we will denote by D. For this type the penalty points are given by

$$pp(type) = \sum_{orders} pp\,(order) + D \sum_{orders} q_1 \cdot q_2^{(t-d+1)^+} \cdot 1_{[0,\infty)}\,(t-d+1)$$

7.2.2. The allocation of capacity

The capacity will be allocated to one or more types of products in several steps. First we determine which of the types is the most important type, of course with a sufficient number of penalty points. If the number of orders for this most important type is less than or equal to the available capacity, then all the orders for this type will be produced. If not, then we will start with the orders with the earliest due date and if orders have the same due date we will start with the orders with the highest priority. The allocation of the capacity stops if all the normal available capacity has been allocated and if the number of periods until the due dates of the remaining orders is larger than some priority-dependent constant. If there are remaining orders for which the number of periods until the due date is not larger than this priority-dependent constant, then extra capacity will be used to produce these orders. In Example 7.1. we will consider orders with two different priorities: normal and urgent orders. For the 'normal' orders we will use extra capacity for all orders with a residual lead time that is less than or equal to one period. For the urgent orders we will use extra capacity for all orders with a residual lead time that is less than or equal to two periods.

First we will consider the situation with one machine. After the capacity allocation for the most important type, there can be some capacity available for the production of another type. If the amount of available capacity is at least *mcl* and if there is another type with a sufficient number of penalty points, then we will allocate capacity to the most important type among the other types. This allocation will be done in exactly the same way as the allocation for the first type. The allocation continues with other types until either the available capacity is less than *mcl*, or until there are no more types with a sufficient number of penalty points or until the multi-type level L has been reached.

In the situation with several identical machines the allocation of capacity can be done in a similar way. If there are several *different* machines the allocation will be made differently. We have to determine the most important type for every machine. Now, the most important types are not simply the types with the highest number of penalty points, but they have to be found by considering the production possibilities on the different machines. This will be worked out in the next subsection. If the most important types have been determined then the available capacity for the important types is allocated in the same way as in the situation with one machine. The capacity allocation for the other types which may be produced in the same period is determined in a slightly different way. This will also be discussed in the next subsection.

7.2.3. Planning on different machines

If the product types are manufactured on several different machines then we have to consider the production possibilities on the different machines. This can be done in several ways. We have chosen for an approach in which we aim for a use of capacity that is more or less equal for all machines. We determine a set of important types in such a way that we can allocate exactly one type to each machine, or if there are not enough types with a sufficient number of penalty points a maximum of one type to each machine. This set is determined in the following way. First we determine the most important type. Of course, this type can be allocated to one of the machines, but this will only be done if we have determined the complete set. After the most important type, we determine the second most important type. Only if it can be manufactured on another machine than the first type, it will be added to the set. We continue determining the next important type and add it to the set if it is possible to produce the new type together with all the types that are already in the set during one period, without allocating more than one type to each of the machines. The test for this possibility is not very difficult. The set is determined if either the number of types in the set is equal to the number of machines or if there are no more types with a sufficient number of penalty points.

If we have determined the set of important types that can be produced in one period then these types are allocated to the different machines. This allocation starts with the types that can only be allocated to one machine, 'removing' the machines to which a type has been allocated, continuing with the types that can only be allocated to two machines and so on. If there are enough types with a sufficient number of penalty points we end up with a situation in which one type has been allocated to each machine. It is possible that after this allocation all capacity has been allocated on some of the machines, whereas there is still capacity available at some of the other machines. We want to use the available capacity as good as possible. Therefore we will try to allocate other types to the machines on which some available capacity is left (at least *mcl*). This will be done in two steps.

In the first step we allocate a type to a machine if all the orders of the type can be produced without exceeding the normal available capacity. We consider the machines in a lexicographical sequence and the types are selected by considering the number of penalty points. In the second step we allocate the types that will be produced one period later, but for which the production can already start in the next period. The choice of the set of important types that will be produced one period later has to be made by making a preliminary allocation of the remaining types to the machines, using the test for the production possibilities. We only consider the types that have not yet been allocated to one of the machines and types for which the expected number of penalty points in the second period is sufficient. On the machines on which the remaining capacity in the firt period is less than the minimum capacity level *mcl* this next type will not be manufactured in the first period. The next type will also not be manufactured if the number of penalty points in the first period is not sufficient. Otherwise we will start the production of the next type and use the capacity in the same way as described for the situation with one machine. The preliminary allocation of the types for the second period will also be used for the determination of the lead times. Every period we will determine the importancy of the different types all over again and we will again determine a set of important types.

7.3. Determination of lead times

In Section 4.4. we have considered a situation with no capacity constraints in which the firm offered the customers lead times. In that section the choice of the lead times was primarily based on the avoidance of set-up costs and holding costs and on the probability that the customers accept the proposed lead time. In this chapter the capacity constraints will have a large influence upon the decision rules for the lead times. In order to avoid high storage costs or high penalty costs the decisions about the lead times will be made according to a preliminary production plan. Using the extended (x,T)–rule that has been described in Section 7.2. we plan the types that will be produced during the first periods. This plan is a preliminary plan and therefore it can be changed in later periods. We suppose that the decision about the production and about the lead times is made at the end of a period. The production planned for the first period can of course not be changed in later periods.

A logical choice is to propose the lead time for an order according to the production period in the preliminary production plan. If we have a maximum lead time of L_{max} we do not have to make decisions about the production in periods later than L_{max} periods. All types, for which there are some newly arrived orders, have to be planned in the preliminary production plan, or otherwise it has to be known that the type will not be produced during the first L_{max} periods. Usually we will have to make a preliminary decision about the production in the first period and in a number of later periods. For the first period the decisions in the production plan will be made according to the rules described in Section 7.2. and these decisions will indeed be performed. For the later periods we want a very simple way of capacity planning and therefore we will use a multi-type level in these periods. Because of the uncertainty of the use of capacity, the lead times for these periods can be different from the delivery times. For the first period we know the exact use of the capacity. Therefore we can give orders for the type(s) that will be produced in this first period a lead time of one period. If the available capacity is not enough for the production of all orders, we assume that the production will be continued in the following period and that the new orders will be given a lead time of two periods. If the available capacity in one of the later periods is not large enough for the production of all the orders, the new orders with the smallest priority will be given a longer lead time. A constant c_L will then be added to the lead time. The value of this decision variable can be roughly the same as the time between set-ups averaged for all types. By limiting the batch size in this way, we can avoid that a large demand for one of the types has a very large influence upon the available capacity for the other types.

In order to make a decision about the types that will be produced in the following periods, we have to consider the expected number of penalty points during those periods. This expected amount will be described in the penalty point function, one for each product type, depending on the period t, containing the sum of the penalty points of all planned orders at time t, the constant C_i multiplied by a number of periods and the sum of the penalty points of the expected future orders until t. We will assume that the future orders will all obtain a lead time that is equal to $t+1$, so the penalty points for these orders will only be q_1, where q_1 may have different values for different types and priorities. Once the type has been allocated to one of the machines, the only penalty points are due to the future orders that arrive after the preliminary production period.

The lead time that the newly arrived orders obtain is that period in which the production of their type is planned, or the maximum lead time if their type is not planned. The type(s) planned to be produced in the first period, will indeed be produced that period, but for later periods, the decisions will be reconsidered every period.

7.3.1. Acceptance of the lead times

The acceptance of the lead time that the firm offers for an order may depend on a lot of elements different from the lead time itself. There may be a long term agreement between the firm and the client, according to which the client accepts all lead times that are less than or equal to an agreed maximum. But also the quality of previous products and the price that is offered will have an influence upon the reaction of the client with respect to a possible withdrawal of an order.

In the examples in this chapter we will consider two simple models. In both models the maximum lead time that is proposed for orders of priority 1 will be 5 periods and for orders of priority 2 it will be 3 periods. In the first model we will assume that the offered lead times will all be accepted. In the second model we will assume a linearly decreasing probability for the acceptance of an order. In Section 4.4. we have considered the same model. The probabilities are slightly different for the different priorities. For orders of priority 1 we assume that the probability that a client accepts a lead time k for an order is given by:

$$A_k = 1.1 - 0.1\,k \qquad k=1,..,5 \qquad (7.3.1)$$

and for orders of priority 2 we have

$$A_k = 1.2 - 0.2\,k \qquad k=1,..,3 \qquad (7.3.2)$$

For both priorities a lead time of 1 period is always accepted.

7.4. Numerical results

In this subsection we will consider two examples, with orders of two different priorities, in which we will use the decision rule described in this chapter. In the performance evaluation we will consider the proposed lead times and the way in which the lead times are met. We will also measure the performance by the revenues and the costs: set-up costs, penalty costs, holding costs and costs for working overtime. We will use two different models for the revenues. In the situation in which all clients accept the lead time that is offered, we will assume that the revenues will depend linearly on the length of the lead times. This is similar to the situation in Section 6.2. in which the revenues depend linearly on the average delivery time. If S_{ij} is the average lead time of an order of type i and priority j and e_{ij} is the average demand per period for these orders and the revenue for an order of type j is R_j, the average revenues per period are given by

$$R = \sum_{i=1}^{N} \sum_{j=1}^{2} R_j \, e_{ij} \, (5 - S_{ij})/4, \tag{7.4.1}$$

where 5 is the maximum lead time and 4 a scaling factor, giving the revenues a similar form as in (7.4.2). In the situation in which not all clients accept the lead time that is offered, the revenues will depend on the number of orders that are indeed placed. This is similar to the situation in Section 4.4. If $\overline{A}_{ij}$ is the average acceptance percentage of orders for type i and priority j, the average revenues per period are given by

$$R = \sum_{i=1}^{N} \sum_{j=1}^{2} R_j \, e_{ij} \, \overline{A}_{ij} \tag{7.4.2}$$

In both examples we will assume a geometrical distribution for the demand, but different from Section 6.2. there is no maximum level for the number of orders during one period for a certain type and priority. We will assume that a set-up takes one unit of time. The multi-type level for the second period and later periods will be set equal to one. In the first example we will not consider a multi-type level for the first period and in the second example the maximum number of types that will be produced on one machine during one period is set equal to three. In both examples we will use the same costs to measure the performance of the rule with a given set of penalty points. We have chosen the set-up costs to be $s = 30$, costs for working overtime $z = 30$ per unit, penalty costs $p_1 = 10$ per period for orders of priority 1 and $p_2 = 30$ per period for orders of priority 2, holding costs $h = 1$ per period and per order and the revenues per order $R_1 = 10$ and $R_2 = 30$.

Example 7.1.

This first example considers the same situation as Example 6.2. with six types of products ($N = 6$). The average demand per period in this example is given by

$$\begin{pmatrix} e_{11} & e_{12} \\ e_{21} & e_{22} \\ e_{31} & e_{32} \\ e_{41} & e_{42} \\ e_{51} & e_{52} \\ e_{61} & e_{62} \end{pmatrix} = \begin{pmatrix} 0.900 & 0.100 \\ 0.750 & 0.250 \\ 2.250 & 0.250 \\ 1.875 & 0.625 \\ 4.500 & 0.500 \\ 3.750 & 1.250 \end{pmatrix}$$

For this example we want to find good values for the parameters q_1 and q_2 in the penalty point functions for the different types and priorities. For the values of q_1 we will start with one of the best choices for the penalty points in Example 6.2. In this example we will use only one value for q_2 which is choosen arbitrarily $q_2=2$ for all types and priorities. This means that starting from the due date period, the penalty points for an order are doubled every period. We start with an arbitrary value for the minimum number of penalty points that is necessary for the production: $x_i=10$ for every type i. The extra points per period are chosen to be $C_i=3$ for every type. These choices have been placed in a matrix XQC_1, where q_{ij} indicates the q_1 value for orders for type i and priority j. The q_2 values are not included in this matrix.

$$XQC_1 = \begin{pmatrix} x_1 & q_{11} & q_{12} & C_1 \\ x_2 & q_{21} & q_{22} & C_2 \\ x_3 & q_{31} & q_{32} & C_3 \\ x_4 & q_{41} & q_{42} & C_4 \\ x_5 & q_{51} & q_{52} & C_5 \\ x_6 & q_{61} & q_{62} & C_6 \end{pmatrix} = \begin{pmatrix} 10 & 2 & 7 & 3 \\ 10 & 2 & 6 & 3 \\ 10 & 1 & 5 & 3 \\ 10 & 1 & 4 & 3 \\ 10 & 1 & 4 & 3 \\ 10 & 1 & 3 & 3 \end{pmatrix}$$

For the calculation of the average amount of work in overtime we will assume that the normal available capacity will be 20 units of time. This is one unit more than the average capacity requirement if we have two set-ups per period. First we perform a simulation with the elements of XQC_1 and the additional elements $c_L=3$, $D=3$ and $mcl=4$. In the second simulation we will make some changes that should promote shorter lead times, however at the cost of more set-ups and more work in overtime. Therefore we replace in XQC_1 the x_i values by 3, giving us the penalty matrix XQC_2 and for the additional elements we choose $D=1$ and $mcl=2$, all leading to shorter lead times. From Table 7.3. we can learn that the lead times indeed have decreased, but with high set-up and overtime costs. Therefore the profit does not increase due to this change. We have tried several other sets of values for the penalty point parameter q_1, but the profit of these sets was never more than one unit higher than the profit of the set XQC_1. Therefore we will not consider the details of these sets. As a last choice we will consider a set XQC_3, which is intended to reduce the number of set-ups and the volume of work in overtime. Therefore we increase the x_i values and also the value of mcl. For the additional elements we have $c_L=3$, $D=2$ and $mcl=5$. The values of the matrix XQC_3 are given by:

$$XQC_3 = \begin{pmatrix} 15 & 2 & 8 & 3 \\ 15 & 2 & 7 & 3 \\ 15 & 1 & 6 & 3 \\ 15 & 1 & 5 & 3 \\ 15 & 1 & 5 & 3 \\ 15 & 1 & 5 & 3 \end{pmatrix}$$

The information of the simulation with the different values for the additional elements and XQC will be summarised in the following tables. In Table 7.1. we will give the percentage of the orders with a certain lead time for the different priorities. In Table 7.2. we will give the differences between the delivery times and the lead times, again in percentages for the different priorities. Finally in Table 7.3. the costs and the profit is given.

$XQCV$-values	XQC_1		XQC_2		XQC_3	
priority	1	2	1	2	1	2
lead time 1	28.2	45.1	53.1	69.4	20.3	41.8
lead time 2	15.2	21.3	13.5	13.6	14.7	19.9
lead time 3	17.2	33.6	8.3	17.0	16.5	38.3
lead time 4	20.9	0.0	12.2	0.0	22.1	0.0
lead time 5	18.4	0.0	12.9	0.0	26.3	0.0

Table 7.1 The lead times in percents in Example 7.1.

$XQCV$-values	XQC_1		XQC_2		XQC_3	
priority	1	2	1	2	1	2
≥2 too late	4.5	2.8	3.6	1.5	7.2	4.4
1 too late	10.2	7.6	11.1	3.4	13.0	8.8
exact	59.8	81.5	69.2	90.3	57.2	78.3
1 too soon	15.7	8.2	7.5	4.8	13.3	8.4
≥2 too soon	9.8	0.0	8.5	0.0	9.3	0.0

Table 7.2 The accuracy of the lead times in percents in Example 7.1.

From these tables we can learn that it is possible in this situation to have very short lead times, on average less than 2 periods for orders of priority 1 and less than 1.5 for orders of priority 2, that it is also possible to have rather accurate lead times and also to have very small overtime costs. However it is difficult to have all these nice

XQCV-values	XQC_1	XQC_2	XQC_3
set-up costs	52.38	79.20	45.48
holding costs	5.77	4.17	5.17
penalty costs	41.72	33.24	60.63
overtime costs	8.40	26.52	5.16
revenues	144.43	177.07	131.07
profit	36.16	33.94	14.63

Table 7.3 The costs and revenues of the penalty sets in Example 7.1.

elements together in this situation in which the machine is occupied more than 90 percent. The choice XQC_1 offers lead times that are not too long and quite accurate, with few set-ups and less than 1.5 percent of work in overtime. The result is that this choice yields the largest profit. From this example we can learn that making a good choice for the penalty points is not very difficult and also that small changes in the penalty points only have little consequences.

Example 7.2.

In this second example we consider an extension of the situation of Example 7.1, with two additional types added to the original six types ($N = 8$). There are three different machines available for the production of these types. Some of the types can be produced on one machine and other types on two machines. The average demand per period for the types in this example is given by

$$\begin{bmatrix} e_{11} & e_{12} \\ e_{21} & e_{22} \\ e_{31} & e_{32} \\ e_{41} & e_{42} \\ e_{51} & e_{52} \\ e_{61} & e_{62} \\ e_{71} & e_{72} \\ e_{81} & e_{82} \end{bmatrix} = \begin{bmatrix} 0.900 & 0.100 \\ 0.750 & 0.250 \\ 1.800 & 0.200 \\ 2.250 & 0.250 \\ 1.875 & 0.625 \\ 2.250 & 0.750 \\ 4.500 & 0.500 \\ 3.750 & 1.250 \end{bmatrix}$$

The production possibilities are given in the matrix B, where $b_{ij} = 0$ if type i cannot be produced on machine j and $b_{ij} = 1$ is type i can be produced on machine j.

$$B = \begin{pmatrix} 1 & 0 & 0 \\ 1 & 1 & 0 \\ 1 & 0 & 0 \\ 0 & 1 & 0 \\ 0 & 1 & 1 \\ 0 & 1 & 0 \\ 0 & 0 & 1 \\ 1 & 0 & 1 \end{pmatrix}$$

In this example the lead times that the firm proposes are accepted with a certain probability A_k. The values of these probabilities have been given in the formulae (7.3.1) and (7.3.2). The normal available capacity on each machine will be 9 units. If no orders would be withdrawn and if we have one set-up on each machine every period, the machines would be occupied more than 90 percent (25 out of 27 units). Due to the withdrawals the real occupancy rate will be around 85 percent. For this example we will first make an arbitrary choice for the penalty points and for the additional elements. We start with $c_L = 3$, $D = 3$ and $mcl = 3$. The q_2 value is chosen arbitrarily as $q_2 = 2$ for all types and priorities. The choices for the other parameters are given in the matrix XQC_4.

$$XQC_4 = \begin{pmatrix} x_1 & q_{11} & q_{12} & C_1 \\ x_2 & q_{21} & q_{22} & C_2 \\ x_3 & q_{31} & q_{32} & C_3 \\ x_4 & q_{41} & q_{42} & C_4 \\ x_5 & q_{51} & q_{52} & C_5 \\ x_6 & q_{61} & q_{62} & C_6 \\ x_7 & q_{71} & q_{72} & C_7 \\ x_8 & q_{81} & q_{82} & C_8 \end{pmatrix} = \begin{pmatrix} 10 & 1 & 5 & 1 \\ 10 & 1 & 5 & 1 \\ 10 & 1 & 5 & 1 \\ 10 & 1 & 5 & 1 \\ 10 & 1 & 5 & 1 \\ 10 & 1 & 5 & 1 \\ 10 & 1 & 5 & 1 \\ 10 & 1 & 5 & 1 \end{pmatrix}$$

From the simulation study of this choice we learned the following things. For the types with a small average demand, a lot of orders of priority 2 have been withdrawn. This implies that the penalty points for orders for these types and this priority have to be increased. We also notice that the lead times for orders of priority 1 are quite long for all types. This can be avoided in two ways: by increasing the C_i values, the number of extra penalty points per period and by decreasing c_L from 3 to 2. For types that can be produced on two machines we have chosen $C_i = 3$ and for types that can only be produced on one machine $C_i = 4$. The value of D and the value of mcl remains unchanged. The choice for XQC_5 is thus given by:

$$XQC_5 = \begin{pmatrix} 10 & 1 & 8 & 4 \\ 10 & 1 & 8 & 3 \\ 10 & 1 & 7 & 4 \\ 10 & 1 & 7 & 4 \\ 10 & 1 & 6 & 3 \\ 10 & 1 & 6 & 4 \\ 10 & 1 & 5 & 4 \\ 10 & 1 & 5 & 3 \end{pmatrix}$$

In the simulation study of this choice we can indeed observe the changes that we have expected. The remaining problem after this simulation is that the types that can be produced on two machines still have a better performance than the other types. Therefore we change the x_i values: types that can be produced on one machine get an x_i value of 8 and the other types a value of 12. We do not change any of the other penalty points or any of the additional elements. The choice for XQC_6 is thus given by:

$$XQC_6 = \begin{bmatrix} 8 & 1 & 8 & 4 \\ 12 & 1 & 8 & 3 \\ 8 & 1 & 7 & 4 \\ 8 & 1 & 7 & 4 \\ 12 & 1 & 6 & 3 \\ 8 & 1 & 6 & 4 \\ 8 & 1 & 5 & 4 \\ 12 & 1 & 5 & 3 \end{bmatrix}$$

The results of the simulation of this choice are already very satisfying. After this simulation we have tried some other choices, with other values for D, c_L and mcl, but none of the changes yielded a higher profit. Finally we have tried some different values for the q_1 values and made some small changes on the C_i and x_i values. The set of values that yields the highest average profit per period is the set given by XQC_7. Compared with XQC_6 we have reduced the number of set-ups a little by increasing x_i for the types that can only be produced on one machine, we have promoted the types 1 and 2, both with a small average demand, by giving them a C_i value of 5 and we have decreased the C_i value for the types 5 and 8, both with a high average demand and both with production possibilities on two machines. The resulting matrix XQC_7 is given by:

$$XQC_7 = \begin{bmatrix} 9 & 1 & 8 & 5 \\ 12 & 1 & 8 & 5 \\ 9 & 1 & 7 & 4 \\ 9 & 1 & 7 & 4 \\ 12 & 1 & 6 & 2 \\ 9 & 1 & 6 & 4 \\ 9 & 1 & 5 & 4 \\ 12 & 1 & 5 & 2 \end{bmatrix}$$

We will summarise the information of the simulation with these four choices for XQC and the corresponding additional elements in the following tables. In Table 7.4. we will give the percentage of the orders for which the lead times have been accepted and the percentages for the different lead times for the accepted orders for the different priorities. In Table 7.5. we will give the differences between the delivery times and the lead times, again in percentages for the different priorities. Finally in Table 7.6. the costs and the profit will be given.

XQCV-values	XQC_4		XQC_5		XQC_6		XQC_7	
priority	1	2	1	2	1	2	1	2
lead time 1	33.2	77.7	31.7	80.7	31.7	79.3	31.2	76.7
lead time 2	21.5	18.3	23.3	17.5	22.1	19.0	22.0	21.0
lead time 3	9.9	4.0	34.0	1.8	34.2	1.6	35.6	2.3
lead time 4	26.2	0.0	10.0	0.0	11.2	0.0	10.4	0.0
lead time 5	9.4	0.0	0.9	0.0	0.7	0.0	0.9	0.0
accepted	81.9	92.1	86.6	93.7	86.5	92.8	85.9	93.9

Table 7.4 The lead times in percents in Example 7.2.

XQCV-values	XQC_4		XQC_5		XQC_6		XQC_7	
priority	1	2	1	2	1	2	1	2
≥2 too late	1.1	0.0	1.0	0.0	1.3	0.0	1.0	0.0
1 too late	7.5	0.7	7.1	1.0	9.0	1.1	8.3	1.3
exact	61.6	98.0	66.0	97.3	64.2	97.6	65.2	97.1
1 too soon	10.8	1.3	16.7	1.7	16.4	1.4	16.6	1.6
≥2 too soon	19.1	0.0	9.3	0.0	9.2	0.0	8.9	0.0

Table 7.5 The accuracy of the lead times in percents in Example 7.2.

In this example we have seen that using the extended (x,T)–rule for the decisions about the production works out very well. Also the use of this rule as a basis for the preliminary production plan on which the lead times can be based works very well. From the results of Table 7.4. we can learn that about 80 percent of the most urgent orders can be produced the next period and that 98 percent can be delivered within two periods. The promised lead times for the most urgent orders are nearly always correct and the number of late orders with priority 1 can be limited within the 10 percent. The percentage of the work that is done outside the normal available capacity is less than 2 percent and this percentage has been reached by only a very simple rule.

XQCV-values	XQC_4	XQC_5	XQC_6	XQC_7
set-up costs	89.88	100.92	96.90	94.63
holding costs	8.45	5.63	5.52	5.44
penalty costs	15.41	15.66	19.49	17.57
overtime costs	19.08	16.50	12.18	13.03
revenues	256.34	266.70	265.52	265.67
profit	123.52	127.99	131.42	134.99

Table 7.6 The costs and revenues of the penalty sets in Example 7.2.

We also have seen that an arbitrary choice for the penalty points and the additional elements yields a performance that is already quite well. Most of the changes that we have considered have been quite obvious. Therefore we expect that a similar approach can be used in a lot of practical situations in which a firm produces to order.

Chapter 8

SUMMARY AND CONCLUSIONS

8.1. Summary and conclusions

In this monograph we have studied several production rules and several rules for proposing due dates, for some different situations. The situations that have been considered share some important elements. The most important one is the element of production to order: the products are manufactured according to customer specifications and they will not be manufactured unless they have been ordered. Other important elements are the set-ups on the machines, which make a clustering of orders for the same product types necessary, the backlogging of late orders and a production process with one bottle-neck.

The objective of this study was to find a simple production rule, which offers possibilities for a simple adaptation to the varying wishes and demands and which can be extended for additional complications. We have concentrated our attention on periodic review models, with usually a time-independent stochastic demand.

In the situation with no capacity restrictions we have considered a number of production rules inspired by two well-known heuristics, the Silver-Meal heuristic and the Wagner-Whitin heuristic, and we have introduced a new, simple rule: the (x,T)–rule. In this (x,T)–rule we do not produce if the number of late orders will be less than x. Otherwise we produce all orders that have to be delivered, according to their due date, within T periods. The x value and the T value are decision variables. If costs for due date deviation and for set-ups are given and if the distribution of the demand is known and stationary, the best pair (x,T) can be determined. The performance of the heuristics and of the (x,T)–rule has been compared with the optimal policy. The difference in average costs per period showed to be very small, whereas the (x,T)–rule is easier to be used and different from the other rules, the (x,T)–rule allows a simple and exact calculation of the average costs.

Because of its performance in uncapacitated situations, the (x,T)–rule seemed to be the best candidate for the simple production rule. Therefore we studied the application of a similar kind of rule in a situation with strict capacity constraints. In this rule, the so-called extended (x,T)–rule, the decisions on production are also based on a largely reduced state space, in which the order state is reduced by a weighting procedure to one simple element: the number of penalty points for this type. For a good use of the capacity, some minor elements have been added to this extended (x,T)–rule. The performance of this rule has been compared with the performance of a more complex rule, inspired by well-known heuristics. This has been done for a situation with strict capacity constraints and also for a situation with possibilities for overtime. In both situations the extended (x,T)–rule performed well compared to the more complex rule.

By assuming such a simple rule we have found several situations in which the performance could be analysed by exact or approximative calculations. This can be done for instance in uncapacitated situations, if the client population depends on the due date deviation or if the clients withdraw their orders if they are dissatisfied by the proposed lead times. The calculation is also possible in situations with capacity restrictions and overtime possibilities, where we assume that only one or only a few product types can be produced during one period.

An extended (x,T)–rule can also be used for proposing lead times for the newly arrived orders. This can be done by making a preliminary production plan, which is based on the expected number of penalty points for the various types in future periods. In the rule for proposing lead times, the basic element is the preliminary production plan, but we can also have additional rules taking into account the capacity constraints, maximum lead times and probabilities for acceptance of the lead times. Some examples of such rules have been discussed in this monograph.

Although the presented (x,T)-rules have been considered for only a limited number of modeled situations, we have obtained the impression that they can be a basis for the solution of a large number of problems, with respect to the production planning and the due date proposals, in situations with production to order. The most important reasons for this belief are the following: the presented rules are simple and clear, they offer a lot of possibilities for direct control, such as the control of lot-sizes and delivery times, and they offer a possibility for an (approximative) analysis in a lot of different situations. With the help of this analysis a lot of changes, for instance a changing market or changing production facilities, can be judged much better.

References

- Anthony R.N. (1965), *"Planning and Control Systems: A Framework for Analysis"*, Harvard University Graduate School of Business Administration, Boston.

- Asmussen S. (1987), *"Applied Probability and Queues"*, Wiley & Sons, New York.

- Baker K.R. (1977), *"An Experimental Study of the Effectiveness of Rolling Schedules in Production Planning"*, Decision Sciences, Vol. 8, pp. 19-27.

- Baker K.R. (1981), *"An Analysis of Terminal Conditions in Rolling Schedules"*, European Journal of Operations Research, Vol. 7, pp. 355-361.

- Bellman R. (1957), *"Dynamic Programming"*, Princeton University Press, Princeton.

- Bemelmans R.P.H.G. (1986), *"The Capacity Aspects of Inventories"*, Springer-Verlag, Berlin, Heidelberg, New York, Tokyo.

- Bitran G., E. Haas and A.C. Hax (1981), *"Hierarchical Production Planning: A Single Stage System"* Operations Research, Vol. 29, pp. 717-743.

- Brown R.G. (1967), *"Decision Rules for Inventory Management"*, Holt, Rinehart and Winston, New York, pp. 29-31.

- Carlson R.C. and D.H. Kropp (1980), *"Inventory Planning using Rolling Production Scheduling"*, Proc. First Int. Symp. on Inventories, Budapest, Hungary.

- Cohen, J.W. and O.J. Boxma (1983), *"Boundary Value Problems in Queuing System Analysis"*, North-Holland, Amsterdam.

- Dellaert N.P. and J. Wessels (1986), *"Production Scheduling with Uncertain Demand"*, Operations Research Proceedings 1985, pp. 247-252.

- Dellaert N.P. (1987), *"Production to Order"*, Memorandum COSOR 87-07, Dept. of Math. and Comput. Sci., Eindhoven University of Technology.

- Dellaert N.P. (1987), *"The Use of the (x,T)-Strategy for Production to Order"*, Memorandum COSOR 87-33, Dept. of Math. and Comput. Sci., Eindhoven University of Technology.

- Dixon P.S. and E.A. Silver (1981), *"A Heuristic Solution Procedure for the Multi Item, Single Level, Limited Capacity, Lotsizing Problem"*, Journal of Operations Management, Vol. 2, pp. 23-39.

- Dogramaci A., J.C. Panayiotopoulos and N.R. Adam (1981), *"The Dynamic Lot-Sizing Problem for Multiple Items under Limited Capacity"*, AIIE Transactions, Vol. 13, pp. 295-303.

- Eisenberg M. (1972), *"Queues with Periodic Service and Changeover Time"*, Operations Research, Vol. 20, pp. 440-451.

- Gorham T. (1968), *"Dynamic Order Quantities"*, Production and Inventory Management, Vol. 9, pp. 75-79

- Graves S.C. (1980), *"The Multi-Product Production Cycling Problem"*, AIIE Transactions, Vol. 12, pp. 233-240.

- Howard R.A. (1960), *"Dynamic Programming and Markov Processes"*, Wiley & Sons, New York.

- Lambrecht M.R. and H. Vanderveken (1979), *"Heuristic Procedures for the Single Operation Multi Item Loading Problem"*, AIIE Transactions, Vol. 11, pp. 319-326.

- Maes J. and L.N. Van Wassenhove (1986), *"A Simple Heuristic for the Multi Item Single Level Capacitated Lotsizing Problem"*, Operations Research Letters, Vol. 4, pp. 265-273.

- Mattsson L., Olhager J., Ovrin P. and B. Rapp (1988), *"Computerization of Manufacturing Planning and Control Systems: Minicomputer-based Systems 1980-1987"*, Paper presented at the Fifth International Working Seminar on Production Economics, Igls, Austria.

- Odoni A. (1969), *"On Finding the Maximal Gain for Markov Decision Processes"*, Operations Research, Vol. 17, pp. 857-860.

- Prabhu N.U. (1965), *"Queues and Inventories"*, Wiley & Sons, New York.

- Ross S.M. (1972), *"Introduction to Probability Models"*, Academic Press, New York and London.

- Silver E.A. and H.C. Meal (1973), *"A Heuristic for Selecting Lot Size Requirements for the Case of a Deterministic Time-Varying Demand Rate and Discrete Opportunities for Replenishment."*, Production and Inventory Management, Vol. 14, pp. 64-74.

- Silver E.A. and R. Peterson (1985), *"Decision Systems For Inventory Management and Production Planning"*, Wiley & Sons, New York.

- Swartz G.B. (1980), *"Polling in a Loop System"*, Journal ACM, Vol. 27, pp. 42-59.

- Van Nunen J.A.E.E. and J. Wessels (1978), *"Multi-Item Lot Size Determination and Scheduling under Capacity Constraints."*, European Journal of Operations Research, Vol. 2, pp. 36-41.

- Vergin R.C. and T.N. Lee (1978), *"Scheduling Rules for the Multiple Product Single Machine System with Stochastic Demand"*, Infor, Vol. 16, pp. 64-73.

- Wagner H.M. and T.M. Whitin (1958), *"Dynamic Version of the Economic Lot Size Model"*, Management Science, Vol. 5, pp. 89-96.

- Watson K.S. (1984), *"Performance Evaluation of Cyclic Service Strategies - A Survey"*, In Performance '84 (E.Gelenbe, editor), North-Holland, Amsterdam, pp. 521-533.

Vol. 236: G. Gandolfo, P.C. Padoan, A Disequilibrium Model of Real and Financial Accumulation in an Open Economy. VI, 172 pages. 1984.

Vol. 237: Misspecification Analysis. Proceedings, 1983. Edited by T.K. Dijkstra. V, 129 pages. 1984.

Vol. 238: W. Domschke, A. Drexl, Location and Layout Planning. IV, 134 pages. 1985.

Vol. 239: Microeconomic Models of Housing Markets. Edited by K. Stahl. VII, 197 pages. 1985.

Vol. 240: Contributions to Operations Research. Proceedings, 1984. Edited by K. Neumann and D. Pallaschke. V, 190 pages. 1985.

Vol. 241: U. Wittmann, Das Konzept rationaler Preiserwartungen. XI, 310 Seiten. 1985.

Vol. 242: Decision Making with Multiple Objectives. Proceedings, 1984. Edited by Y.Y. Haimes and V. Chankong. XI, 571 pages. 1985.

Vol. 243: Integer Programming and Related Areas. A Classified Bibliography 1981–1984. Edited by R. von Randow. XX, 386 pages. 1985.

Vol. 244: Advances in Equilibrium Theory. Proceedings, 1984. Edited by C.D. Aliprantis, O. Burkinshaw and N.J. Rothman. II, 235 pages. 1985.

Vol. 245: J.E.M. Wilhelm, Arbitrage Theory. VII, 114 pages. 1985.

Vol. 246: P.W. Otter, Dynamic Feature Space Modelling, Filtering and Self-Tuning Control of Stochastic Systems. XIV, 177 pages. 1985.

Vol. 247: Optimization and Discrete Choice in Urban Systems. Proceedings, 1983. Edited by B.G. Hutchinson, P. Nijkamp and M. Batty. VI, 371 pages. 1985.

Vol. 248: Plural Rationality and Interactive Decision Processes. Proceedings, 1984. Edited by M. Grauer, M. Thompson and A.P. Wierzbicki. VI, 354 pages. 1985.

Vol. 249: Spatial Price Equilibrium: Advances in Theory, Computation and Application. Proceedings, 1984. Edited by P. T. Harker. VII, 277 pages. 1985.

Vol. 250: M. Roubens, Ph. Vincke, Preference Modelling. VIII, 94 pages. 1985.

Vol. 251: Input-Output Modeling. Proceedings, 1984. Edited by A. Smyshlyaev. VI, 261 pages. 1985.

Vol. 252: A. Birolini, On the Use of Stochastic Processes in Modeling Reliability Problems. VI, 105 pages. 1985.

Vol. 253: C. Withagen, Economic Theory and International Trade in Natural Exhaustible Resources. VI, 172 pages. 1985.

Vol. 254: S. Müller, Arbitrage Pricing of Contingent Claims. VIII, 151 pages. 1985.

Vol. 255: Nondifferentiable Optimization: Motivations and Applications. Proceedings, 1984. Edited by V.F. Demyanov and D. Pallaschke. VI, 350 pages. 1985.

Vol. 256: Convexity and Duality in Optimization. Proceedings, 1984. Edited by J. Ponstein. V, 142 pages. 1985.

Vol. 257: Dynamics of Macrosystems. Proceedings, 1984. Edited by J.-P. Aubin, D. Saari and K. Sigmund. VI, 280 pages. 1985.

Vol. 258: H. Funke, Eine allgemeine Theorie der Polypol- und Oligopolpreisbildung. III, 237 pages. 1985.

Vol. 259: Infinite Programming. Proceedings, 1984. Edited by E.J. Anderson and A.B. Philpott. XIV, 244 pages. 1985.

Vol. 260: H.-J. Kruse, Degeneracy Graphs and the Neighbourhood Problem. VIII, 128 pages. 1986.

Vol. 261: Th.R. Gulledge, Jr., N.K. Womer, The Economics of Made-to-Order Production. VI, 134 pages. 1986.

Vol. 262: H.U. Buhl, A Neo-Classical Theory of Distribution and Wealth. V, 146 pages. 1986.

Vol. 263: M. Schäfer, Resource Extraction and Market Structure. XI, 154 pages. 1986.

Vol. 264: Models of Economic Dynamics. Proceedings, 1983. Edited by H.F. Sonnenschein. VII, 212 pages. 1986.

Vol. 265: Dynamic Games and Applications in Economics. Edited by T. Başar. IX, 288 pages. 1986.

Vol. 266: Multi-Stage Production Planning and Inventory Control. Edited by S. Axsäter, Ch. Schneeweiss and E. Silver. V, 264 pages. 1986.

Vol. 267: R. Bemelmans, The Capacity Aspect of Inventories. IX, 165 pages. 1986.

Vol. 268: V. Firchau, Information Evaluation in Capital Markets. VII, 103 pages. 1986.

Vol. 269: A. Borglin, H. Keiding, Optimality in Infinite Horizon Economies. VI, 180 pages. 1986.

Vol. 270: Technological Change, Employment and Spatial Dynamics. Proceedings 1985. Edited by P. Nijkamp. VII, 466 pages. 1986.

Vol. 271: C. Hildreth, The Cowles Commission in Chicago, 1939–1955. V, 176 pages. 1986.

Vol. 272: G. Clemenz, Credit Markets with Asymmetric Information. VIII, 212 pages. 1986.

Vol. 273: Large-Scale Modelling and Interactive Decision Analysis. Proceedings, 1985. Edited by G. Fandel, M. Grauer, A. Kurzhanski and A.P. Wierzbicki. VII, 363 pages. 1986.

Vol. 274: W.K. Klein Haneveld, Duality in Stochastic Linear and Dynamic Programming. VII, 295 pages. 1986.

Vol. 275: Competition, Instability, and Nonlinear Cycles. Proceedings, 1985. Edited by W. Semmler. XII, 340 pages. 1986.

Vol. 276: M.R. Baye, D.A. Black, Consumer Behavior, Cost of Living Measures, and the Income Tax. VII, 119 pages. 1986.

Vol. 277: Studies in Austrian Capital Theory, Investment and Time. Edited by M. Faber. VI, 317 pages. 1986.

Vol. 278: W.E. Diewert, The Measurement of the Economic Benefits of Infrastructure Services. V, 202 pages. 1986.

Vol. 279: H.-J. Büttler, G. Frei and B. Schips, Estimation of Disequilibrium Models. VI, 114 pages. 1986.

Vol. 280: H.T. Lau, Combinatorial Heuristic Algorithms with FORTRAN. VII, 126 pages. 1986.

Vol. 281: Ch.-L. Hwang, M.-J. Lin, Group Decision Making under Multiple Criteria. XI, 400 pages. 1987.

Vol. 282: K. Schittkowski, More Test Examples for Nonlinear Programming Codes. V, 261 pages. 1987.

Vol. 283: G. Gabisch, H.-W. Lorenz, Business Cycle Theory. VII, 229 pages. 1987.

Vol. 284: H. Lütkepohl, Forecasting Aggregated Vector ARMA Processes. X, 323 pages. 1987.

Vol. 285: Toward Interactive and Intelligent Decision Support Systems. Volume 1. Proceedings, 1986. Edited by Y. Sawaragi, K. Inoue and H. Nakayama. XII, 445 pages. 1987.

Vol. 286: Toward Interactive and Intelligent Decision Support Systems. Volume 2. Proceedings, 1986. Edited by Y. Sawaragi, K. Inoue and H. Nakayama. XII, 450 pages. 1987.

Vol. 287: Dynamical Systems. Proceedings, 1985. Edited by A.B. Kurzhanski and K. Sigmund. VI, 215 pages. 1987.

Vol. 288: G.D. Rudebusch, The Estimation of Macroeconomic Disequilibrium Models with Regime Classification Information. VII, 128 pages. 1987.

Vol. 289: B.R. Meijboom, Planning in Decentralized Firms. X, 168 pages. 1987.

Vol. 290: D.A. Carlson, A. Haurie, Infinite Horizon Optimal Control. XI, 254 pages. 1987.

Vol. 291: N. Takahashi, Design of Adaptive Organizations. VI, 140 pages. 1987.

Vol. 292: I. Tchijov, L. Tomaszewicz (Eds.), Input-Output Modeling. Proceedings, 1985. VI, 195 pages. 1987.

Vol. 293: D. Batten, J. Casti, B. Johansson (Eds.), Economic Evolution and Structural Adjustment. Proceedings, 1985. VI, 382 pages. 1987.

Vol. 294: J. Jahn, W. Krabs (Eds.), Recent Advances and Historical Development of Vector Optimization. VII, 405 pages. 1987.

Vol. 295: H. Meister, The Purification Problem for Constrained Games with Incomplete Information. X, 127 pages. 1987.

Vol. 296: A. Börsch-Supan, Econometric Analysis of Discrete Choice. VIII, 211 pages. 1987.

Vol. 297: V. Fedorov, H. Läuter (Eds.), Model-Oriented Data Analysis. Proceedings, 1987. VI, 239 pages. 1988.

Vol. 298: S.H. Chew, Q. Zheng, Integral Global Optimization. VII, 179 pages. 1988.

Vol. 299: K. Marti, Descent Directions and Efficient Solutions in Discretely Distributed Stochastic Programs. XIV, 178 pages. 1988.

Vol. 300: U. Derigs, Programming in Networks and Graphs. XI, 315 pages. 1988.

Vol. 301: J. Kacprzyk, M. Roubens (Eds.), Non-Conventional Preference Relations in Decision Making. VII, 155 pages. 1988.

Vol. 302: H.A. Eiselt, G. Pederzoli (Eds.), Advances in Optimization and Control. Proceedings, 1986. VIII, 372 pages. 1988.

Vol. 303: F.X. Diebold, Empirical Modeling of Exchange Rate Dynamics. VII, 143 pages. 1988.

Vol. 304: A. Kurzhanski, K. Neumann, D. Pallaschke (Eds.), Optimization, Parallel Processing and Applications. Proceedings, 1987. VI, 292 pages. 1988.

Vol. 305: G.-J.C.Th. van Schijndel, Dynamic Firm and Investor Behaviour under Progressive Personal Taxation. X, 215 pages. 1988.

Vol. 306: Ch. Klein, A Static Microeconomic Model of Pure Competition. VIII, 139 pages. 1988.

Vol. 307: T.K. Dijkstra (Ed.), On Model Uncertainty and its Statistical Implications. VII, 138 pages. 1988.

Vol. 308: J.R. Daduna, A. Wren (Eds.), Computer-Aided Transit Scheduling. VIII, 339 pages. 1988.

Vol. 309: G. Ricci, K. Velupillai (Eds.), Growth Cycles and Multisectoral Economics: the Goodwin Tradition. III, 126 pages. 1988.

Vol. 310: J. Kacprzyk, M. Fedrizzi (Eds.), Combining Fuzzy Imprecision with Probabilistic Uncertainty in Decision Making. IX, 399 pages. 1988.

Vol. 311: R. Färe, Fundamentals of Production Theory. IX, 163 pages. 1988.

Vol. 312: J. Krishnakumar, Estimation of Simultaneous Equation Models with Error Components Structure. X, 357 pages. 1988.

Vol. 313: W. Jammernegg, Sequential Binary Investment Decisions. VI, 156 pages. 1988.

Vol. 314: R. Tietz, W. Albers, R. Selten (Eds.), Bounded Rational Behavior in Experimental Games and Markets. VI, 368 pages. 1988.

Vol. 315: I. Orishimo, G.J.D. Hewings, P. Nijkamp (Eds.), Information Technology: Social and Spatial Perspectives. Proceedings, 1986. VI, 268 pages. 1988.

Vol. 316: R.L. Basmann, D.J. Slottje, K. Hayes, J.D. Johnson, D.J. Molina, The Generalized Fechner-Thurstone Direct Utility Function and Some of its Uses. VIII, 159 pages. 1988.

Vol. 317: L. Bianco, A. La Bella (Eds.), Freight Transport Planning and Logistics. Proceedings, 1987. X, 568 pages. 1988.

Vol. 318: T. Doup, Simplicial Algorithms on the Simplotope. VIII, 262 pages. 1988.

Vol. 319: D.T. Luc, Theory of Vector Optimization. VIII, 173 pages. 1989.

Vol. 320: D. van der Wijst, Financial Structure in Small Business. VII, 181 pages. 1989.

Vol. 321: M. Di Matteo, R.M. Goodwin, A. Vercelli (Eds.), Technological and Social Factors in Long Term Fluctuations. Proceedings. IX, 442 pages. 1989.

Vol. 322: T. Kollintzas (Ed.), The Rational Expectations Equilibrium Inventory Model. XI, 269 pages. 1989.

Vol. 323: M.B.M. de Koster, Capacity Oriented Analysis and Design of Production Systems. XII, 245 pages. 1989.

Vol. 324: I.M. Bomze, B.M. Pötscher, Game Theoretical Foundations of Evolutionary Stability. VI, 145 pages. 1989.

Vol. 325: P. Ferri, E. Greenberg, The Labor Market and Business Cycle Theories. X, 183 pages. 1989.

Vol. 326: Ch. Sauer, Alternative Theories of Output, Unemployment, and Inflation in Germany: 1960–1985. XIII, 206 pages. 1989.

Vol. 327: M. Tawada, Production Structure and International Trade. V, 132 pages. 1989.

Vol. 328: W. Güth, B. Kalkofen, Unique Solutions for Strategic Games. VII, 200 pages. 1989.

Vol. 329: G. Tillmann, Equity, Incentives, and Taxation. VI, 132 pages. 1989.

Vol. 330: P.M. Kort, Optimal Dynamic Investment Policies of a Value Maximizing Firm. VII, 185 pages. 1989.

Vol. 331: A. Lewandowski, A.P. Wierzbicki (Eds.), Aspiration Based Decision Support Systems. X, 400 pages. 1989.

Vol. 332: T.R. Gulledge, Jr., L.A. Litteral (Eds.), Cost Analysis Applications of Economics and Operations Research. Proceedings. VII, 422 pages. 1989.

Vol. 333: N. Dellaert, Production to Order. VII, 158 pages. 1989.

GPSR Compliance

The European Union's (EU) General Product Safety Regulation (GPSR) is a set of rules that requires consumer products to be safe and our obligations to ensure this.

If you have any concerns about our products, you can contact us on ProductSafety@springernature.com

In case Publisher is established outside the EU, the EU authorized representative is:

Springer Nature Customer Service Center GmbH
Europaplatz 3
69115 Heidelberg, Germany

Batch number: 10844827

Printed by Printforce, the Netherlands